Fire Detection and Suppression Systems

Fourth Edition

Michael J. Sturzenbecker
Project Manager/Senior Editor

Barbara Adams
Senior Editor

Elkie Burnside
Graduate Associate

ifsta

Validated by the International Fire Service Training Association

Published by
Fire Protection Publications • Oklahoma State University

RECYCLABLE

The International Fire Service Training Association

The International Fire Service Training Association (IFSTA) was established in 1934 as a *nonprofit educational association of fire fighting personnel who are dedicated to upgrading fire fighting techniques and safety through training*. To carry out the mission of IFSTA, Fire Protection Publications was established as an entity of Oklahoma State University. Fire Protection Publications' primary function is to publish and disseminate training texts as proposed and validated by IFSTA. As a secondary function, Fire Protection Publications researches, acquires, produces, and markets high-quality learning and teaching aids as consistent with IFSTA's mission.

The IFSTA Validation Conference is held the second full week in July. Committees of technical experts meet and work at the conference addressing the current standards of the National Fire Protection Association® and other standard-making groups as applicable. The Validation Conference brings together individuals from several related and allied fields, such as:

- Key fire department executives and training officers
- Educators from colleges and universities
- Representatives from governmental agencies
- Delegates of firefighter associations and industrial organizations

Committee members are not paid nor are they reimbursed for their expenses by IFSTA or Fire Protection Publications. They participate because of commitment to the fire service and its future through training. Being on a committee is prestigious in the fire service community, and committee members are acknowledged leaders in their fields. This unique feature provides a close relationship between the International Fire Service Training Association and fire protection agencies, which helps to correlate the efforts of all concerned.

IFSTA manuals are now the official teaching texts of most of the states and provinces of North America. Additionally, numerous U.S. and Canadian government agencies as well as other English-speaking countries have officially accepted the IFSTA manuals.

ISBN 978-0-87939-398-4   *Library of Congress Control Number: 2010940147*

Fourth Edition, First Printing, November 2010   *Printed in the United States of America*

10   9   8   7   6   5   4   3   2

If you need additional information concerning the International Fire Service Training Association (IFSTA) or Fire Protection Publications, contact:

Customer Service, Fire Protection Publications, Oklahoma State University
930 North Willis, Stillwater, OK 74078-8045
800-654-4055   Fax: 405-744-8204

For assistance with training materials, to recommend material for inclusion in an IFSTA manual, or to ask questions or comment on manual content, contact:

Editorial Department, Fire Protection Publications, Oklahoma State University
930 North Willis, Stillwater, OK 74078-8045
405-744-4111   Fax: 405-744-4112   E-mail: editors@osufpp.org

Glossary

Glossary

A

Acceptance Test — Preservice test on fire detection and/or suppression systems after installation to ensure that the system operates as intended.

Adjustable Fog Nozzle — Nozzle designed to allow the discharge pattern to be adjusted from straight stream to full fan fog; suitable for applying water, wet water, or foam solution. Some adjustable fog nozzles allow the rate of flow to be adjusted as well.

Aeration — Introduction of air into a foam solution to create bubbles that result in finished foam.

Aerosol — Form of mist characterized by highly respirable, minute liquid particles.

Agent — Generic term used for materials that are used to extinguish fires.

Air-Aspirating Foam Nozzle — Foam nozzle designed to provide the aeration required to make the highest quality foam possible; most effective appliance for the generation of low-expansion foam.

Alarm Check Valve — Type of check valve installed in the riser of an automatic sprinkler system that transmits a water-flow alarm when the water flow in the system lifts the valve clapper.

Alarm Circuit — (1) Electrical circuit connecting two points in a fire alarm system; for example, from the signal device to the fire station, from the central alarm center to all fire stations, or from the sending device to the audible alarm services. (2) The circuit on a fire alarm system that connects the alarm initiating devices, such as the smoke detectors to the fire alarm control panel.

Aqueous Film Forming Foam (AFFF) — Synthetic foam concentrate that, when combined with water, can form a complete vapor barrier over fuel spills and fires and is a highly effective extinguishing and blanketing agent on hydrocarbon fuels.

Atmospheric Pressure — Force exerted by the atmosphere at the surface of the earth due to the weight of air. Atmospheric pressure at sea level is about 14.7 psi (101 kpa) and can be measured as 760 mm of mercury on a barometer. Atmospheric pressure increases as elevation decreases, and decreases as elevation increases.

Authority Having Jurisdiction (AHJ) — Term used in codes and standards to identify the legal entity, such as a building or fire official, that has the statutory authority to enforce a code and to approve or require equipment; may be a unit of a local, state, or federal government, depending on where the work occurs. In the insurance industry it may refer to an insurance rating bureau or an insurance company inspection department.

Automatic Sprinkler System — System of water pipes, discharge nozzles, and control valves designed to activate during fires by automatically discharging enough water to control or extinguish a fire.

Auxiliary Alarm System — System that connects the protected property with the fire department alarm communications center by a municipal master fire alarm box or over a dedicated telephone line.

B

Balanced Pressure Proportioner — A foam concentrate proportioner that operates in tandem with a fire water pump to ensure a proper foam concentrate-to-water mixture.

Ball Valve — Valve having a ball-shaped internal component with a hole through its center that permits water to flow through when aligned with the waterway.

Bimetallic — Strip or disk composed of two different metals that are bonded together; used in heat-detection equipment.

Branch Line — Those pipes in an automatic sprinkler system to which the sprinklers are directly attached.

British Thermal Unit (Btu) — Amount of heat energy required to raise the temperature of 1 lb of water 1°F. 1 Btu = 1.055 kilo joules (kJ).

C

Cavitation — Condition in which vacuum pockets form due to localized regions of low pressure at the vanes in the impeller of a centrifugal pump and cause vibrations, loss of efficiency, and possible damage to the impeller.

Central Station System — Alarm system that functions through a constantly attended location (central station) operated by an alarm company. Alarm signals from the protected property are received in the central station and are then retransmitted by trained personnel to the fire department alarm communications center.

Centrifugal Pump — Pump with one or more impellers that rotate and utilize centrifugal force to move the water. Most modern fire pumps are of this type.

Check Valve — Automatic valve that permits liquid flow in only one direction. For example, the inline valve that prevents water from flowing into a foam concentrate container when the nozzle is turned off or there is a kink in the hoseline.

Churn — Rotation of a centrifugal pump impeller when no discharge ports are open so that no water flows through the pump.

Circulation Relief Valve — Small relief valve that opens and provides enough water flow into and out of the pump to prevent the pump from overheating when it is operating at churn against a closed system.

Clean Agent — Fire suppression media that leaves little or no residue when used.

Clean-Agent Fire-Extinguishing System — System that uses special extinguishing agents that leave little or no residue.

Code — A collection of rules and regulations enacted by a legislative body to become law in a particular jurisdiction.

Cold Smoke — Smoke from a fire that lacks any substantial heat.

Combination Detector — Alarm-initiating device that is capable of detecting an abnormal condition by more than one means. The most common combination detector is the fixed-temperature/rate-of-rise heat detector.

Combination System — Water supply system that is a combination of both gravity and direct pumping systems. It is the most common type of municipal water supply system.

Compartmentation Systems — Series of barriers designed to keep flames, smoke, and heat from spreading from one room or floor to another; barriers may be doors, extra walls or partitions, fire-stopping materials inside walls or other concealed spaces, or floors.

Complex Loop — Piping system that is characterized by one or more of the following: more than one inflow point, more than one outflow point, and/or more than two paths between inflow and outflow points.

Compound Gauge — Pressure gauge capable of measuring above and below atmospheric pressure usually found on the intake of a fire pump and measures the intake pressure on a fire pump.

Compressed Air Foam System (CAFS) — Generic term used to describe a high-energy foam-generation system consisting of an air compressor (or other air source), a water pump, and foam solution that injects air into the foam solution before it enters a hoseline.

Consensus Standard — Rules, principles, or measures that are established though agreement of members of the standards-setting organization.

Controller — Electric control panel used to switch a fire pump on and off and to control its operation.

D

Dead-End Main — Water main that is not looped and in which water can flow in only one direction.

Deflector — Part of the sprinkler assembly that creates the discharge pattern of the water.

Deluge Sprinkler System — Fire-suppression system that consists of piping and open sprinklers. A fire detection system is used to activate the water or foam control valve. When the system activates, the extinguishing agent expels from all sprinkler heads in the designated area.

Deluge Valve — Automatic valve used to control water to a deluge sprinkler system.

Drain Valve — Valve that allows piping to drain when pressure is relieved in the pipe.

Driver — Engine or motor used to turn a pump.

Dry-Barrel Hydrant — Fire hydrant that has its operating valve at the water main rather than in the barrel of the hydrant. When operating properly, there is no water in the barrel of the hydrant when it is not in use. These hydrants are used in areas where freezing could occur.

Chapter Summary

Table of Contents

List of Tables

Preface

The fourth edition of IFSTA's **Fire Detection and Suppression Systems** is designed to serve as an introductory text to fire protection systems. In doing so, this manual will familiarize fire department and industrial fire protection personnel with the various types of fire protection systems found in public and private buildings. This text is also designed to meet Fire and Emergency Services Higher Education (FESHE) course outcomes for the Fire Protection Systems core course.

Acknowledgement and special thanks are extended to the members of the IFSTA validating committee who contributed their time, wisdom, and knowledge to the development of this manual.

IFSTA Fire Detection and Suppression Systems
Fourth Edition
Validation Committee

Chair
Bradd Clark
Fire Chief
Owasso Fire Department
Owasso, Oklahoma

Vice Chair
Martin King
Assistant Chief
West Allis Fire Department
West Allis, Wisconsin

Secretary
Brett Lacey
Fire Marshal
Colorado Springs Fire Department
Colorado Springs, Colorado

Committee Members

Andrew Barr
Senior Fire Systems Engineer
McKinney Fire Department
McKinney, Texas

Larry Collins
Professor/Chair
Department of Safety, Security, and
  Emergency Management
Eastern Kentucky University
Richmond, Kentucky

Samuel Feltner
Fire Protection Inspector
Wright-Patterson Air Force Base Fire Department
Wright-Patterson Air Force Base, Ohio

Jason Finney
Fire Protection Design Engineer
Babcock & Wilcox Technical Services Y-12, LLC
Knoxville, Tennessee

Tim Frankenberg
Captain
Washington Fire Department
Washington, Missouri

Dave Hanneman
Fire Chief
Chula Vista Fire Department
Chula Vista, California

Committee Members (continued)

Jerry Howell
Fire Marshal
Lombard Fire Department
Lombard, Illinois

Brent Meisenheimer
Firefighter
Austin Fire Department
Austin, Texas

Jimbo Schifiliti
President
Fire Safety Consultants, Inc.
Elgin, Illinois

Josh Stefancic
Staff Lieutenant
Tarpon Springs Fire Rescue
Tarpon Springs, Florida

Ed Steiner
Director of Building Services and
   Fire Code Enforcement
City of Edmond
Edmond, Oklahoma

Scott Strassburg
Fire Prevention Officer
Madison Fire Department
Madison, Wisconsin

Much appreciation is given to the following individuals and organizations for contributing information, photographs, and technical assistance instrumental in the development of this manual:

Gaylen Brevik
Castle Rock Firefighting Consulting

Chicago Fire Department
Chicago, Illinois

Ben DaSilva
Chief Engineer
Juhl Unit Owners' Association
Las Vegas, Nevada

Doddy Photography

Fire Service Training
Oklahoma State University

Tim Frankenberg, CFPS
Fire Product Manager
Potter Electric Signal Company
St. Louis, Missouri

Goodfellow Air Force Base Fire Department
Goodfellow Air Force Base, Texas

Tom Hughes
Ingalls Fire Department
Ingalls, Oklahoma

Los Angeles Fire Department – ISTS
Los Angeles, California

Floyd Luinstra
School of Fire Protection and Safety Technology
Oklahoma State University

Rich Mahaney

Ron Moore
McKinney Fire Department
McKinney, Texas

North Las Vegas Fire Department
North Las Vegas, Nevada

Oklahoma State University
Risk Management and Environmental
   Health and Safety

Sonny Pelayo, P.E.
Engineer
Austin Fire Department
Austin, Texas

Mike Porowski

Rolf Jensen & Associates

Sand Springs Fire Department
Sand Springs, Oklahoma

Steven Shacklett, CFPS
Fire Inspector
Las Vegas Fire and Rescue
Las Vegas, Nevada

Josh Stefancic
Staff Lieutenant
Tarpon Springs Fire Rescue
Tarpon Springs, Florida

Stillwater Fire Department
Stillwater, OK

Texas Fire Protection Specialists
Carrollton, Texas

Underwriters Laboratories, Inc.

United States Department of Defense

Last, but certainly not least, gratitude is extended to the following members of the Fire Protection Publications staff whose contributions made the final publication of this manual possible.

IFSTA Fire Detection and Suppression Systems Project Team

Project Manager
Mike Sturzenbecker, Senior Editor

Technical Writer
Beverley Walker
Hall County Fire Services
Gainesville, Georgia

Editors/Proofreaders
Barbara Adams, Senior Editor
Libby Hieber, Senior Editor
Lynne Murnane, Senior Editor
Elkie Burnside, Graduate Associate

Production Manager
Ann Moffat, Coordinator, Publications Production

Illustrators and Layout Designers
Missy Hannan, Senior Graphic Designer
Ruth Mudroch, Senior Graphic Designer

Curriculum Development
Beth Ann Fulgenzi, Curriculum Developer
Elkie Burnside, Curriculum Developer
Andrea Haken, Curriculum Developer

Photographer
Jeff Fortney, Senior Editor

Technical Reviewer
Glen Albright
Fire Prevention Specialist
Ventura City Fire Department
Ventura, California

Library Researchers
Susan Walker, Librarian

Editorial Staff
Ed Kirtley, IFSTA/Curriculum
  Projects Coordinator
Tara Gladden, Editorial Assistant
Gabriel Ramirez, Research Assistant

The IFSTA Executive Board at the time of validation of the **Fire Detection and Suppression Systems** manual was as follows:

IFSTA Executive Board

Chair
Jeffrey Morrissette
Commission on Fire Prevention and Control
Windsor Locks, Connecticut

Vice Chair
Paul Valentine
Nexus Engineering
Oakbrook Terrace, Illinois

Executive Director
Mike Wieder
Fire Protection Publications
Stillwater, Oklahoma

Board Members

Stephen Ashbrock
Madeira & Indian Hill Fire Department
Cincinnati, Ohio

Steve Austin
Cumberland Valley Vol. Firemen's Assn.
Newark, Delaware

Roxanne Bercik
Los Angeles Fire Department
Los Angeles, California

Mary Cameli
Mesa Fire Department
Mesa, Arizona

Bradd Clark
Owasso Fire Department
Owasso, Oklahoma

Dennis Compton
Mesa, Arizona

Frank Cotton
Memphis Fire Department
Memphis, Tennessee

George Dunkel
Scappoose, Oregon

John Hoglund
Maryland Fire and Rescue Institute
College Park, Maryland

Wes Kitchel
Santa Rosa Fire Department
Santa Rosa, California

Brett Lacey
Colorado Springs Fire Department
Colorado Springs, Colorado

Ernest Mitchell
Cerritos, California

Lori Moore-Merrell
International Association of Fire Fighters
Washington, District of Columbia

Introduction

Introduction Contents

Introduction

Fire detection and suppression systems are installed and used in a multitude of different occupancies. These systems are designed to detect the presence of fire or products of combustion, alert occupants and fire department personnel of the condition, and suppress or extinguish the fire. Components of these systems are constantly changing and improving due to advances in technology.

IFSTA's *Fire Detection and Suppression Systems*, 4th edition is written to meet the Fire and Emergency Services Higher Education (FESHE) model course outcomes for the Fire Protection Systems core course. With this objective in mind, two new chapters have been added to the manual: Introduction to Fire Detection and Suppression Systems and Smoke Management Systems.

Chapter 1, Introduction to Fire Detection and Suppression Systems provides historical background information on these systems and a brief description of historical major loss fires that influenced fire code development. In addition, major codes, standards, and legislation that have influenced the development and improvement of fire detection and suppression systems are discussed.

Chapter 3, Smoke Management Systems provides an overview of how smoke behaves and is dealt with in structures. The chapter begins with an introduction to smoke behavior and addresses key concepts in smoke management. The remainder of the chapter addresses different passive and active smoke ventilation methods.

The 4th edition of IFSTA's *Fire Detection and Suppression Systems* manual is intended to be an introductory learning resource for those who may be responsible for the design, installation, inspection and maintenance of these systems. In addition, the manual is also intended to be a resource for emergency services personnel who may respond to incidents in protected premises. The manual has been completely revised to reflect the most current equipment, technology, and practices in the field.

Purpose and Scope

The purpose of the fourth edition of *Fire Detection and Suppression Systems* is to familiarize fire service and other interested personnel with the components, design, maintenance, operation, testing and inspection of common fire detection and suppression systems. This manual is not intended to substitute for training as a fire inspector or for systems design, maintenance, and inspection personnel, but can serve as a valuable resource for personnel engaged in these activities.

The scope of this manual is to provide up-to-date information on fire detection and suppression systems. The audience for this manual includes college students, fire service personnel, industrial fire protection personnel, building construction personnel, and all others seeking information on fire detection and suppression systems. It contains an introduction to fire detection and suppression systems and information on water supply, automatic sprinkler systems including residential sprinkler systems, standpipe and hose systems, fire pumps, and special extinguishing systems. It also addresses fire alarm and detection systems, smoke management systems, and portable fire extinguishers. Information on the design, operation, maintenance, and inspection of these systems and equipment is provided.

Key Information

Various types of information in this manual are given in shaded boxes marked by symbols or icons. See the following definitions:

Safety Alert

Provides additional emphasis on matters of safety.

Case History

A case history analyzes an event. It can describe its development, action taken, investigation results, and lessons learned. Illustrations can be included.

Pump Controller — Electric control panel used to switch a fire pump on and off and to control its operation.

A key term is designed to emphasize key concepts, technical terms, or ideas that firefighters need to know. They are listed at the beginning of each chapter and the definition is placed in the margin for easy reference.

Three key signal words are found in the text: **WARNING, CAUTION,** and **NOTE.** Definitions and examples of each are as follows:

- **WARNING** indicates information that could result in death or serious injury to fire and emergency services personnel. See the following example:

WARNING!
Because roof supports almost always run perpendicular to the outside walls, firefighters should never walk diagonally across the roof of a burning building.

- **CAUTION** indicates important information or data that fire and emergency service responders need to be aware of in order to perform their duties safely. See the following example:

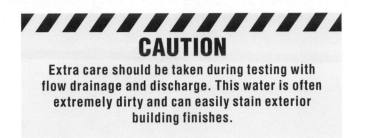

CAUTION

Extra care should be taken during testing with flow drainage and discharge. This water is often extremely dirty and can easily stain exterior building finishes.

- **NOTE** indicates important operational information that helps explain why a particular recommendation is given or describes optional methods for certain procedures. See the following example:

NOTE: Class D ratings are never assigned to any type of multipurpose extinguisher.

Overview of Fire Detection and Suppression Systems

Divider page photo courtesy of Tarpon Springs (FL) Fire Rescue

Key Terms

FESHE Outcomes

Fire and Emergency Services Higher Education (FESHE) Outcomes: Fire Protection Systems

1. Explain the benefits of fire protection systems in various types of structures.
5. Review residential and commercial sprinkler legislation.

Overview of Fire Detection and Suppression Systems

Learning Objectives

After reading this chapter, students will be able to:

1. Define the terms *fire detection system* and *fire suppression system*.

2. Summarize the historical development of fire detection and suppression systems.

3. Describe the various types of commonly used fire detection and suppression systems.

4. Summarize the role of codes and standards in the design, installation, and maintenance of fire detection and suppression systems.

5. Describe the role and benefits of residential and commercial fire detection and suppression systems in protecting life and property.

6. Summarize the impact of fire detection and suppression systems on firefighter safety and fire suppression operations.

Chapter 1
Overview of Fire Detection and Suppression Systems

Case History

A fire occurred in a large single-family residence in Arizona in 2008. The house was unoccupied but was equipped with a fire detection and alarm system as well as a residential sprinkler system. A signal was sent to an alarm-monitoring service who notified the fire department. Upon arrival, firefighters quickly located the fire and found that it was contained in the area of origin by the residential sprinkler system. Because these systems worked as intended, the property owner was able to resume use of the residence after only minor repairs.

Fire prevention has long been recognized as a responsibility of fire and emergency services. In its truest form, fire prevention is the elimination of all hostile fires through education, engineering, and enforcement. It can also involve the **mitigation** of loss or limiting property loss or injury. This can be accomplished through product regulation, design, or redesign.

Fire prevention also involves behavior modification that prevents or mitigates the threat from hostile fires. Changing behavior can be accomplished through education and by specifically targeting high-risk audiences. Recognizing that fire prevention efforts will never be 100 percent effective, fire detection and suppression systems are designed to mitigate damage to both life and property when fires occur.

Fire detection and suppression systems can be part of the built environment, acting automatically upon discovery of a fire. They can also be used manually, as with fire extinguishers or manual alarm systems. This chapter serves as an introduction to fire detection and suppression systems, including the types, history, and benefits of systems. In addition, the chapter discusses important codes and standards that impact the installation and use of fire detection and suppression systems.

Mitigate — To cause to become less harsh or hostile; to make less severe, intense, or painful; to alleviate.

Types of Fire Protection Systems

Fire protection systems can be divided into two broad categories: fire detection and fire suppression. **Fire detection systems** are those that detect hazardous conditions. The early detection of a fire and the signaling of an appropriate alarm remains the most significant factor in preventing large losses due to fire. **Fire suppression systems** are those that control or contain hazardous conditions. Many of these systems are water-based. Working together, fire detection and suppression systems decrease loss through prompt notification and early mitigation and control of the fire **(Figure 1.1)**.

Fire protection can be either passive or active. Passive fire protection can be a part of building construction such as a concrete firewall or sprayed-on structural fire protection **(Figure 1.2)**. These systems are passive in that they perform with no outside intervention or mechanical support. Passive fire protection cannot warn occupants of the dangers of an unwanted fire or suppress a growing fire.

Active fire detection and suppression systems are those that activate in some fashion during a fire to sound an alarm or contain a small fire. These systems require some type of outside intervention or mechanical support such as electricity or a water supply. Active systems, when activated, are designed to change the course and outcome of a fire in a building **(Figure 1.3)**. These systems work to the benefit of the building owner, its occupants, and the responding fire department personnel.

Figure 1.1 Multiple components work together to provide for quality fire detection and suppression systems.

Figure 1.2 Sprayed-on structural fire stopping is an example of passive fire protection.

Figure 1.3 Active fire protection devices, such as sprinklers, work to change the course and outcome of a fire.

History of Fire Detection and Suppression Systems

When looking at fire protection, the concept of detection and suppression systems is deeply rooted in history. The first fire alarm system was installed around 1851. This was a municipal alarm system that used the telegraph to notify fire departments of a fire. The first commercially successful automatic fire sprinkler system was patented in 1872. In 1902, George Andrew Darby, an electrical engineer, patented the first heat indicator, which sounded a fire alarm when activated. This early device paved the way for today's battery-powered smoke alarms, which were introduced in 1969 by Kenneth House and Randolph Smith.

There have been many catastrophic fires throughout history that have served to increase the promotion and use of fire detection and suppression systems. Many of these events prompted changes in codes and regulations concerning the design, installation, and maintenance of fire protection systems. These devastating events have become benchmarks in the history of fire detection and suppression systems and provide much evidence as to the life-saving benefits of these systems. Notable fires include the following:

- *Iroquois Theater Fire, Chicago, Illinois, 1903* — This fire claimed the lives of 602 people. The scenery in the playhouse was made of canvas and painted in highly flammable oil paints. A hot stage light ignited a velvet curtain, causing the backdrops to burst into flames. There were no automatic sprinklers for the stage, and the stage's fire curtain did not close properly. There was no emergency lighting, the smoke and heat vents for the stage were not functional, and many exit doors were either locked or did not open in the proper direction.

- *Triangle Shirtwaist Factory, New York City, 1911* — This company was located in a high-rise building, occupying the 8th, 9th, and 10th floors. An employee tossed a cigarette into a bin containing scrap material, causing it to ignite immediately. While employees on the 8th and 10th floors were able to escape due to early discovery and notification, the 250 employees on the 9th floor were not as fortunate. They were not informed of the fire and the dangerous situation it presented until the structure was well involved. The fire occurred at 4:45 p.m., around the time when the business was closing for the day. As was the custom to deter theft, security guards had already locked one of the two exit doors on the 9th floor. Some individuals were able to escape from the unlocked exit, but it was quickly made inaccessible by the fire. There were 145 lives lost in this fire.

- *Cocoanut Grove Nightclub, Boston, Massachusetts, 1942* — A fire in this nightclub claimed the lives of 492 individuals. The club's walls were covered with paper palm tree decorations that ignited when someone struck a match. The nightclub was filled to twice its capacity, having approximately 1,000 occupants at the time the fire erupted. As with the Iroquois Theater fire, there were no automatic sprinklers, the exit doors did not swing in the proper direction, many doors and windows were sealed shut, and the primary exit was a revolving door.

- *Winecoff Hotel, Atlanta, Georgia, 1946* — This hotel was the scene of the deadliest hotel fire in United States history, causing 119 fatalities. The building had only one exit stairway, which became impassable early in the fire's development. Doors had been propped open, and there was no fire suppression system in the building. The building was not equipped with a fire alarm system, so there was no way to notify the hotel's occupants when the fire started in the early morning hours.

- *Katie Jane Memorial Home, Warrenton, Missouri, 1957* — This event was one of the deadliest nursing home fires in American history. Of the 149 elderly residents, 71 perished in the fire, and one died later at the hospital. This fire was believed to have started due to faulty wiring in the ceiling or in a wall. There was no sprinkler system in the facility, nor was there an alarm system, manual or automatic. The only means of escape for occupants were unenclosed interior stairways that emptied into the main building, and there were no outside fire stairs or slides.

- *Our Lady of Angels School, Chicago, Illinois, 1958* — The deadliest school fire in United States history started in a basement trash can in the stairwell, and the structure's wooden staircase was quickly engulfed in flames. The location of the fire blocked the escape route for the occupants of the second floor. The exit corridors had combustible walls and ceilings, and once again there were no automatic fire sprinklers. In addition, there was no automatic fire alarm system, the stairwell was not enclosed, and there was a delay in notification of the fire department. This fire resulted in the deaths of 92 school children and 3 nuns.

- *Beverly Hills Supper Club, Southgate, Kentucky, 1977* — This incident was another nightclub fire that impacted codes and regulations concerning fire protection systems. It was Memorial Day weekend in 1977, and the building was packed with over 3,000 patrons. There was inadequate egress and exit identification, no fire suppression system and no alarm system. The fire started in the Zebra Room, an unoccupied room, and was believed to be electrical in nature. The fire burned for some time before being discovered and quickly spread to other parts of the building. The fire took the lives of

165 individuals, many of whom were enjoying a show in the Caberet Room at the opposite end of the building. It was noted that employees attempted to fight the fire prior to notifying the fire department or the building's occupants.

- **MGM Grand Hotel, Las Vegas, Nevada, 1980** — A fire at this hotel caused 87 fatalities and almost 700 injuries. While the hotel portion of the building was protected by a sprinkler system, there were no sprinklers in the casino area. The hotel contained many unprotected vertical shafts and had openings that allowed smoke to enter and fill the exit stairwells. Once the occupants entered the stairwells, they were unable to return to their floor or room due to automatic locking doors.

- **Station Nightclub, West Warwick, Rhode Island, 2003** — One of the most recent notable fires in the United States took the lives of 100 patrons of this nightclub. Pyrotechnics used during a band performance ignited soundproofing foam at the back of the stage. There were no automatic sprinklers. Individuals panicked and failed to use secondary exit routes. There was a rush for the front door, which became jammed, and most of the victims died trying to exit at the front of the club **(Figure 1.4)**.

Figure 1.4 Photo of the Station Club fire in Rhode Island. *Courtesy of Mike Porowski.*

The necessity of fire detection and suppression systems is made obvious by the large loss of life in these events as well as the need for early notification of a building's occupants and the fire department. As a result of these events, codes and regulations concerning both active and passive fire protection systems have been adopted, modified, or more strictly enforced.

In addition to these historical benchmarks, there are two early reports that have influenced the requirements for fire detection and suppression systems. These reports were prepared by groups that came together to address the unacceptable fire problem in America, and both promoted the use of fire protection devices for life safety and loss prevention during fires.

In 1947, President Harry S. Truman commissioned The President's Conference on Fire Prevention. This group was brought together to review the fire problem in the United States and to identify deficiencies in America's fire service. The purpose of the group was to determine ways to increase the level of awareness about the fire problem and to increase the "work of fire safety in every community." As to fire detection and suppression systems, the report stated that all municipal fire prevention ordinances should be combined into one Fire Prevention Code. The report recommended better building design, the use of technology for prevention, and an increased focus on prevention education. Specifically, the report recommended the widespread use of fire extinguishing equipment, hoses, standpipes, automatic sprinklers, and alarm systems in buildings.

The current concept of the *Three E's of Prevention* — education, engineering, and enforcement — came as a result of this conference report. Those involved in fire prevention practices realized that in order to reduce the number of fire

Figure 1.5 Station tours are a method of fire prevention education.

deaths and related injuries, all three components must be implemented. As a result, fire prevention programs achieved more successful outcomes **(Figure 1.5)**.

President Richard Nixon convened the National Commission on Fire Prevention and Control in the 1970s. This commission produced the famous report entitled *America Burning*, which has had a significant impact on the fire service, fire prevention, and fire protection in the United States. Overall recommendations included the improvement of fire protection features of buildings. Specifically, this report identified the need for "automatic fire extinguishing systems in every high-rise building and every low-rise building where people congregate." The report recommended that economic incentives be provided for built-in fire protection.

This report resulted in the passage of the Federal Fire Prevention and Control Act of 1974. The impact of this publication has been enormous in the areas of fire prevention, fire protection, fire service training, and public education. Since that time, the requirements for fire suppression and detection systems in commercial structures have been incorporated and expanded. Presently, in many jurisdictions there are codes and regulations addressing requirements for such systems in one- and two-family dwellings.

Overview of Fire Protection Systems

There are a variety of fire detection and suppression systems, based upon their function and purpose. These systems can be divided into five basic categories:

1. *Automatic fire detection and alarm systems* — These systems are activated by smoke, high temperatures, or radiant heat and alert the occupants in the building and/or the fire department to the fire. These systems do not slow the growth of the fire or reduce the amount of smoke produced.

2. *Automatic fire suppression systems* — These systems actually work to suppress the fire hazard by applying a fire-suppressing medium to the fire. There is no need for human intervention for their effectiveness. This reduces the hazard to the occupants and to the building and its contents.

3. *Manual fire alarm systems* — These systems require activation by individuals upon discovery of a fire. Devices such as manual pull stations are common components of manual systems **(Figure 1.6)**.

Figure 1.6 This pull station is an example of a manual fire alarm system.

Figure 1.7 Class III standpipes are incorporated into structures as manual fire suppression systems.

4. *Manual fire suppression systems* — This category includes standpipes (both vertical and horizontal), fire hoses, and fire extinguishers. These mechanisms require an individual to operate them in order to perform their function; that is, they are not automatic (**Figure 1.7**).

5. *Smoke control/exhaust systems* — These systems mechanically remove smoke and other products of combustion from the structure. When unwanted fires occur, removing or controlling the movement of smoke is important in certain types of occupancies or structures. Smoke removal increases the survivability of victims, enhances fire fighting efforts, and reduces the likelihood of fire spread in the structure.

Standards and Codes for Detection and Suppression Systems

As indicated earlier, standards and codes have traditionally been developed and adopted in response to historical tragic events. Both are used to establish minimum requirements for the design, construction, and use of buildings, structures, and facilities and provide for the type and extent of detection and suppression systems.

Standards and codes establish the minimum level of safety that should be present in a structure. These minimums must be adopted by the authority having jurisdiction (AHJ) and are enforced by building and fire officials. While the terms are often used interchangeably, they are not identical.

A **standard** is a set of principles, protocols, or procedures that is developed by committees through a consensus process. Standards describe how to do something or provide a minimum set of principles that should be followed.

Standard — Criterion documents that are developed to serve as models or examples of desired performance or behaviors and that contain requirements and specifications outlining minimum levels of performance, protection, or construction.

Code — A collection of rules and regulations enacted by a legislative body to become law in a particular jurisdiction.

A **code** is a collection of rules and regulations enacted by a legislative body to become law in a particular jurisdiction. The code can be composed of standards, either in their entirety or in part. Simply stated, a code is a law or body of regulations for a political subdivision such as a municipality or county.

The main difference between a standard and a code is the level of enforcement that they provide. Standards provide a model or desired level of performance, and codes make them enforceable. Basically, a code is a law that may be based on or incorporate a standard. A standard only becomes law when it is legally adopted by a jurisdiction or is included as part of a code **(Figure 1.8)**.

Standards and codes are often created by organizations who specialize in their development. In order to be enforceable by law, standards and codes must be adopted by the AHJ. The following information addresses development and adoption of standards and codes.

Comparison Between Codes and Standards

Codes	Standards
Address one broad topic	Address one specific topic
Based on requirements described in standards	Establish design, behavior, and installation criteria
May be amended when adopted by the AHJ	Become law only when adopted by the AHJ
Have the force of law when adopted	Developed through a consensus process

Figure 1.8 Although the terms codes and standards are sometimes used interchangeably, they have different meanings.

Transcription — Method by which an AHJ adopts a code in whole to become a new regulation.

Reference — Method by which an AHJ refers to a code in a regulation and states that the code is legally enforceable.

Consensus Standard — Rules, principles, or measures that are established though agreement of members of the standards-setting organization.

Adoption of Standards and Codes

Building and fire codes can be adopted in two ways. The first is by **transcription**, where the entire code is copied into a regulation. The other method is by **reference**, where the regulation states that the referenced code is legally enforceable as part of the fire and life safety regulations. Codes can be adopted in whole or in part and can be amended by the jurisdiction once adopted.

Standards attempt to obtain consistency in design, practice, and materials and can provide a guide of practices and designs that have proven to be successful. Many standards are **consensus standards**, which mean that a group of experts has developed and agreed upon the standard before it is adopted.

Before codes were developed, there was a wide variety of interpretations and even confusion for requirements. Model codes were developed to provide agreed-upon requirements for areas such as fire and life safety or electrical equipment. These codes are similar to consensus standards in that a group agrees upon the code before it is adopted into the model.

Standards- and Codes-Developing Organizations

There are a variety of organizations that publish consensus standards and model codes. The most familiar is the National Fire Protection Association® (NFPA®). The NFPA® publishes the majority of the consensus standards used in the United States and Canada. Since the 1880s, the NFPA® has published a variety of standards and other documents dealing with fire prevention and building safety. Many of these standards identify the various fire detection and suppression systems that are required for specific occupancies or buildings.

The International Code Council (ICC) is a membership association dedicated to building safety and fire prevention. It was established in 1994 as a nonprofit organization that is dedicated to developing a single set of comprehensive national model construction codes. The ICC develops the codes used to construct residential and commercial buildings, including homes and schools. The founders of the ICC are Building Officials and Code Administrators International, Inc. (BOCA), International Conference of Building Officials (ICBO), and Southern Building Code Congress International, Inc. (SBCCI). These three organizations had previously developed separate sets of model codes that were used throughout the United States. Most U.S. cities, counties, and states that adopt codes choose the International Codes developed by the ICC.

ASTM International (originally known as the American Society for Testing and Materials) is one of the largest standard developing organizations in the world and is considered to be a trusted source for technical standards for materials, products, systems, and services. ASTM International develops testing processes that other testing organizations use in the development of safety products.

Underwriters Laboratories Inc. (UL) is an independent, not-for-profit product certification organization that has been testing products and writing standards for safety since 1894. UL evaluates thousands of products, components, materials, and systems annually **(Figure 1.9)**. UL has developed over 800 standards for safety. Some of these relate directly to fire and life safety such as smoke detectors for fire alarm systems and dry chemical fire extinguishers.

FM Global (formerly Factory Mutual) is an independent organization that works to provide property improvements for business owners to prevent risk. The organization works in partnership with businesses and industries to reduce property damage through state-of-the-art property-loss prevention research and engineering and comprehensive insurance products. FM Global provides product certification and testing services through an approval process and specification testing.

Figure 1.9 UL is one of several nationally recognized testing laboratories. In this test, structural materials are placed in a furnace to determine their fire resistance. *Courtesy of Underwriters Laboratories, Inc.*

There are nationally recognized testing laboratories (NRTLs) that are recognized by the U.S. Occupational Safety and Health Administration (OSHA) as having the capability to provide product safety testing and certification services to the manufacturers of a wide range of products. The testing and certifications are based on widely accepted product safety standards and are often issued by the American National Standards Institute (ANSI). These testing agencies must meet the certification requirements of OSHA's Directorate of Technical Support and Emergency Management. Both UL and FM Global are certified as NRTLs through OSHA. Other smaller testing laboratories include Southwest Research Institute (SwRI), Wyle Laboratories (WL), the Canadian Standards Association (CSA), and many others. OSHA provides a complete list of these agencies on its web site.

ANSI coordinates the development and use of voluntary consensus standards in the United States. ANSI is also actively involved in accrediting programs that assess conformance to standards in businesses, including the International Organization for Standardization (ISO) 9000 and ISO 14000 management systems. ANSI standards may be cross-referenced between NFPA® and OSHA documents.

It is from this wide array of sources that jurisdictions base their building and fire codes. These codes designate the degree or level of protection required in a building based upon its occupancy, construction type, and other designated variables.

Benefits of Fire Detection and Suppression Systems

The influence of fire detection and suppression systems on life safety cannot be overstated. The positive effects for both building occupants and firefighters are readily apparent by an examination of historical data. In each of the fires discussed previously, the investigations revealed the lack of fire alarm systems and fire suppression systems as a major contributing factor in the deadly results.

According to estimates by the NFPA® and the United States Fire Administration (USFA), U.S. home usage of smoke alarms rose from less than 10 percent in 1975 to at least 95 percent in 2000. At the same time, the number of home fire deaths was cut nearly in half. The relationship between improved use of smoke alarms and lower death rates due to fire is not a coincidence. Because of this drastic reduction in fire deaths, the home smoke alarm is widely credited as the greatest success story in fire safety in the last part of the 20th century **(Figure 1.10)**.

Figure 1.10 Fire departments often install smoke alarms for free in their communities. *Courtesy of Tarpon Springs (FL) Fire Rescue.*

The NFPA® states that a properly installed and maintained automatic sprinkler system will reduce the average property loss from fire by one-half to two-thirds. According to the USFA, property losses are 85 percent less in buildings that are protected with fire sprinklers compared to those without sprinklers. The combination of automatic sprinklers and early warning systems in all buildings and residences could reduce overall injuries, loss of life, and property damage by at least 50 percent.

The USFA has promoted research, development, testing, and demonstrations of residential fire sprinkler systems for more than 30 years. USFA research regarding residential fire sprinkler systems clearly shows that these systems can do the following:

- Save the lives of building occupants.
- Save the lives of firefighters called to respond to a home fire.
- Significantly offset the risk of premature building collapse posed to firefighters by lightweight construction components when they are involved by fire.
- Substantially reduce property loss caused by a fire.

The Fire and Life Safety Section of the International Association of Fire Chiefs (IAFC) has stated priorities for the promotion of fire and life safety. These priorities include reducing death from structure fires to zero and limiting property damage from structures to the area of origin as well as reducing firefighter fireground deaths to zero. In pursuit of these priorities, the IAFC advocates complete automatic fire sprinkler protection in all new and occupied construction, including one- and two-family dwellings. It also advocates fire sprinkler retrofits of existing high-rises, institutional, and other high-risk/high-consequence occupancies. In addition, the IAFC recommends the education of all firefighters on the benefits of fire sprinkler protection in reducing firefighter deaths.

In addition to protecting civilians and their properties, there are also benefits to emergency responders. In its work to reduce the number of firefighter deaths and injuries, the National Fallen Firefighters Foundation, through its Everyone Goes Home® Firefighter Life Safety Initiatives Program, states that "advocacy must be strengthened for the enforcement of codes and the installation of home fire sprinklers."

Summary

Fire detection and suppression systems are of vital importance in the goal of fire prevention. While these systems may not be able to prevent a fire from occurring, they certainly work effectively and efficiently to prevent loss of life or large property loss. History demonstrates the seriousness of the fire problem and the tragic toll it can take on a community or fire department. Fire detection and suppression systems allow fire professionals to achieve their goal of life safety and property protection.

Review Questions

1. What is the difference between a fire detection system and a fire suppression system?
2. What are three notable fires in the history of the United States?
3. What report was produced by the National Commission on Fire Prevention and Control?
4. What are the basic categories of fire detection and suppression systems?
5. What is the difference between a standard and a code?
6. What are the two ways that building and fire codes can be adopted?
7. What organization publishes the majority of consensus standards used in the United States and Canada?
8. What are three standards- and codes- developing organizations?
9. What are two benefits of fire detection and suppression systems?
10. How can firefighters benefit from residential sprinkler systems?

Fire Detection and Alarm Systems

Chapter Contents

chapter 2

Key Terms

FESHE Outcomes

Fire and Emergency Services Higher Education (FESHE) Outcomes: Fire Protection Systems

7. Explain the basic components of a fire alarm system.

8. Identify the different types of detectors and explain how they detect fire.

Fire Detection and Alarm Systems

Learning Objectives

After reading this chapter, students will be able to:

1. Describe components of a fire detection and alarm system.

2. Discuss factors that affect signaling system requirements and types of specialty signals.

3. Describe protected premises alarm systems.

4. Distinguish among types of supervising station alarm systems.

5. Describe types of emergency communications systems.

6. Discuss types of alarm initiating devices.

7. Discuss types of fixed-temperature heat detectors.

8. Distinguish among types of rate-of-rise heat detectors.

9. Distinguish among types of smoke detectors.

10. Discuss flame detectors, fire-gas detectors, and other detection devices.

11. Discuss acceptance testing of fire detection and alarm systems.

12. Discuss service testing and periodic inspection of fire detection and alarm systems.

13. Summarize general inspection and testing requirements and timetable guidelines for various types of systems.

14. Discuss fire detection and alarm system record keeping.

Chapter 2
Fire Detection and Alarm Systems

Case History

On July 21, 2007 an alarm monitoring company received an automatic fire alarm from a residence in California. Through an intercom system integrated into the alarm, the monitoring company was able to make contact with the residents and determine that there was indeed a fire in the residence. The representative from the monitoring company then attempted to report the alarm to the fire dispatch center by calling a nonemergency line. Because the representative failed to state there was a confirmed fire in the residence, they were placed on hold by dispatchers to attend to higher priority calls. Due to this confusion, nearly 10 minutes elapsed between the activation of the fire alarm and the time the first fire unit went en route.

By the time fire units arrived on scene, the fire had grown significantly. Firefighters made entry and attempted to rescue two victims inside the structure. Sadly, two firefighters died during this rescue attempt. While there were a number of factors that contributed to this unfortunate incident, the significant delay between the time the alarm was activated and when fire units were dispatched greatly compounded the situation.

The early detection of a fire and the signaling of an appropriate alarm remain two of the most significant factors in preventing the loss of life and property due to fire. History has proven that a delay in the detection of a hazardous condition and the communication of an alarm can lead to injuries, fatalities, and property loss. Properly installed and maintained fire detection and alarm systems increase the survivability of occupants and emergency responders while also reducing the risk of a large-loss incident.

The combination of fire detection and alarm systems is just one facet of the fire protection systems found in all property types. These systems can be complicated and highly technical. Fire detection and alarm systems that are installed in residential occupancies are usually less complicated, although they operate on the same principles and have many of the same components.

This chapter provides information on the basic components and types of fire detection and alarm systems, both manual and automatic. In addition, the inspection and testing of these systems will be discussed as well as the importance of record keeping and timetables.

Fire Alarm System Components

Modern detection and signaling systems vary in complexity from those that are very simple to systems that incorporate advanced detection and signaling equipment. Such systems can be designed and installed by qualified individuals as determined by the authority having jurisdiction (AHJ). The design, installation, and approval of a fire detection and alarm system may also require acceptance by multiple regulatory agencies before it can be used as an in-service unit **(Figure 2.1)**.

Figure 2.1 Fire detection and alarm system plans must usually be approved by the AHJ prior to installation and use.

A nationally recognized testing laboratory (NRTL), such as Underwriters Laboratories Inc. (UL), FM Global, or Intertek (ETL Listed Mark), should test the components of a system to ensure operational reliability. In addition, the installation of the system should conform to the applicable provisions of NFPA® 70 *National Electrical Code®*, NFPA® 72 *National Fire Alarm and Signaling Code*, and local codes and ordinances.

Several basic components are found in various fire detection and alarm systems. These components include fire alarm control panels (FACPs), primary and secondary power supplies, initiating devices, and notification appliances. The components of fire detection and alarm systems are discussed in more detail in the following sections.

Fire Alarm Control Panels

Fire Alarm Control Panel (FACP) — A system component that receives input from automatic and manual fire alarm devices and may provide power to detection devices or communication devices.

The **fire alarm control panel (FACP)** contains the electronics that supervise and monitor the fire alarm system **(Figure 2.2)**. The FACP basically serves as the *brain* for the alarm system. It receives signals from alarm initiating devices, processes the signals, and produces output signals. Power and fire alarm circuits are connected directly into this panel. In addition, the power boosters and power supplies for the notification system are considered to be part of the FACP.

Controls for the system are located in the FACP. The FACP can also perform other functions such as the control of a remote annunciator, the operation of relays that capture and recall elevators, or public address and mass notifications.

Primary Power Supply

Figure 2.2 FACPs supervise and monitor the fire alarm system.

The primary electrical power supply usually is obtained from the building's main connection to the local utility provider **(Figure 2.3)**. In rare instances where electrical service is unavailable or unreliable, primary power supply can be provided by an engine-driven generator. If such a generator is used, either a trained operator must be on duty 24 hours a day or the system must contain multiple engine-driven generators. One of these generators must always be set for automatic starting. The primary power supply must be supervised and must signal an alarm if the power supply is interrupted.

Secondary Power Supply

A secondary power supply must be provided for the detection and alarm system. This is done to ensure that the system will be operational even if the main power supply fails. The secondary power supply must be capable of making the system fully operational within a specified time period. The time period requirements for secondary power operation capabilities vary and can be found in NFPA® 72. Secondary power sources can consist of batteries with chargers, engine-driven generators with a storage battery, or multiple engine-driven generators, of which one must be set for automatic starting **(Figure 2.4)**.

Initiating Devices

Initiating devices are the manual and automatic devices that are activated or sense the presence of the products of combustion or other hazardous conditions and then send a signal to the FACP. The initiating devices may be either connected to the FACP by the hard-wire system or radio-controlled over a special frequency. Initiating devices include but are not limited to manual pull stations, heat detectors, smoke detectors, flame detectors, waterflow devices, tamper switches, combination detectors, and other specialty suppression systems **(Figure 2.5)**. Initiating devices will be discussed in more detail later in this chapter.

Figure 2.3 In most instances, power for the system is obtained from the local utility provider.

Figure 2.4 Generators are a common secondary power source. *Courtesy of the McKinney (TX) Fire Department.*

Figure 2.5 Manual pull stations are one of several types of initiating devices.

Notification Appliances

Once an initiating device activates and sends a signal to the FACP, the signal is processed by the control unit and the appropriate action is taken. Local notification devices include bells, buzzers, horns, recorded voice messages, strobe lights, speakers, and other warning appliances **(Figure 2.6, p. 24)**.

Audible notification signaling appliances are the most common types of alarm-signaling systems used for signaling a fire alarm in a structure. Depending on the design of the system, the local alarm may either sound only

Figure 2.6 Speaker notification appliances allow for voice instructions to be given to occupants of the structure.

Figure 2.7 This combination appliance incorporates both audible and visual signals.

Figure 2.8 Bed shakers are integrated with the detection device in the occupancy to alert those who are deaf or hard-of-hearing to an emergency while sleeping.

Smoke Damper — Device that restricts the flow of smoke through an air-handling system; usually activated by the building's fire alarm signaling system.

in the area of the activated detector or sound in the entire facility. Notification appliances fall under the following categories which can be used in any combination **(Figure 2.7)**:

- *Audible* — Approved sounding devices such as horns, bells, or speakers that indicate a fire or emergency condition.

- *Visual* — Approved lighting devices such as strobes or flashing lights that indicate a fire or emergency condition.

- *Textual* — Visual text or symbols indicating a fire or emergency condition.

- *Tactile* — Indication of a fire or emergency condition through sense of touch or vibration **(Figure 2.8)**.

Off-Site Reporting

There are several different methods that monitored systems use to send transmissions from the facility to the monitoring station. These methods include the following:

- *Digital alarm communicator transmitter (DACT)* — Provides transmission through two separate phone lines (one at a time) to provide a redundant method of signal transmission to a monitoring station.

- *Digital alarm radio system (DARS)* — Sends a radio signal from a digital radio transmitter located at the protected premises to a monitoring station.

- *Cellular* — Uses cellular phone technology to transmit a signal from the protected premises to a monitoring station.

- *Internet protocol* — Transmits a signal from the protected premises to a monitoring station via an approved Internet connection.

- *City tie/polarity reversal* — Uses a hardwired system that communicates a signal from the protected premises directly to the emergency services telecommunications center or other approved location.

Additional Alarm System Functions

Building codes have special requirements for some types of occupancies in the event of a possible fire condition. In these cases, the fire detection and alarm system can be designed to perform the following special functions:

- Turn off the heating, ventilating, and air-conditioning (HVAC) system for smoke control.

- Close **smoke dampers** and/or fire doors **(Figure 2.9)**.

- Pressurize stairwells for evacuation purposes.

- Override control of elevators and prevent them from operating on the fire floor.

- Automatically return the elevator to a designated evacuation floor.

- Operate heat and smoke vents.

- Activate special fire-extinguishing systems such as preaction and deluge sprinkler systems or a variety of special-agent fire-extinguishing systems.

Figure 2.9 Smoke dampers can be closed remotely by the system.

Figure 2.10 Places of assembly, such as movie theaters, often require more substantial detection and alarm systems.

Types of Signaling Systems

Signaling systems range from the very simple to the very complex. A simple system may only sound a local evacuation alarm; whereas a more complex system may sound a local alarm, activate building services, and notify fire and security agencies to respond. The type of system required depends upon the type of occupancy of the building and is affected by the following factors:

- Level of life safety hazard **(Figure 2.10)**
- Structural features of the building
- Level of hazard presented by the contents of the building
- Availability of fire-suppression resources
- State and local code requirements

For many years, **smoke detectors** (now called *conventional smoke detectors*) were neither analog nor addressable. Since they were non-analog, there were only two operating states: normal and alarm. Because they were not addressable, conventional smoke detectors were unable to provide a specific location to the FACP. Today, modern smoke detectors may have analog capabilities, which means they can monitor the conditions in a protected space. In addition, addressable smoke detectors are now available, which will communicate the specific location of the device that is activated to the FACP.

Smoke Detector — Alarm-initiating device designed to actuate when visible or invisible products of combustion (other than fire gases) are present in the room or space where the unit is installed.

Smoke Alarms vs. Smoke Detectors

The terms smoke alarm and smoke detector are often used interchangeably. While this is common practice, it is technically incorrect. Smoke alarms are the devices typically installed in residential occupancies. These devices combine a smoke detector with a local notification appliance. When activated, smoke alarms emit an audible alarm to notify occupants of the presence of smoke.

Smoke detectors differ from smoke alarms in that they do not include a local notification appliance. When activated, smoke detectors send a signal to a FACP or a similar device. The FACP then initiates the alarm to notify occupants.

Fire alarm systems are equipped with a number of specialty signals, depending on the type and nature of the alarm they are reporting. An alarm signal is a warning of a fire emergency or dangerous condition that demands immediate attention. A **supervisory signal** is one that is used to indicate an off-normal condition of the complete fire protection system. Supervisory signals also include a returned-to-normal signal, meaning that the condition has been resolved. These are used to monitor the integrity of the fire protection features of the system. A **trouble signal** indicates a problem with a monitored circuit or component of the fire alarm system or the power supply for the system. This signal must be clearly audible to attract attention so that corrective action can be taken.

Note: A trouble signal indicates a problem with the fire alarm itself. A supervisory alarm indicates a problem with an accessory of the fire alarm system.

There are several system types that may be encountered by fire personnel. Individuals should be able to recognize each type of system and understand how each system operates. This is important when performing inspections or conducting preincident planning. Major types of systems include the following:

- Protective premises (Local)
- Supervising station alarm
  - Auxiliary
  - Proprietary
  - Central station
  - Remote receiving
- Emergency communications systems

Both emergency communications systems and parallel telephone systems may be found in buildings with certain occupancies or building types. Mass notification systems are a special type of emergency communications system that may be found as a part of a building's alarm system to provide specific and detailed instructions to a building's occupants. NFPA® 72 contains the requirements for all fire alarm and protective signaling systems and should be consulted for further information. Major systems are discussed in the following sections.

Protected Premises (Local)

A **protected premises system** or *local alarm system* is designed to provide notification to building occupants only on the immediate premises. Where these systems are allowed, there are no provisions for automatic off-site reporting. The local system can be activated by manual means, such as a pull station, or by automatic devices such as smoke detectors. A local system may also be capable of annunciating a supervisory or trouble condition to ensure that service interruptions do not go unnoticed. Local systems can be designed to activate the auxiliary services described later in this chapter. There are three basic types of local alarm systems: noncoded, zoned/annunciated, and addressable alarms.

Presignal Alarms

Presignal alarms are unique systems that may be employed in locations such as hospitals where greater assistance is needed to help occupants evacuate in a safe and orderly manner. The system initially responds with a presignal that alerts emergency personnel before the general occupancy is notified. This presignal is usually a discreet signal that is recognizable only by personnel who are familiar with the system. The presignal may be a recorded message over an intercom, a soft alarm signal, or a pager notification. The presignal provides emergency personnel with an opportunity to assist the general occupancy in evacuation. Depending on the policies of the occupancy and local code requirements, emergency personnel may elect to handle the incident without sounding a general alarm. Personnel may elect to sound the general alarm after investigating the problem, or the general alarm will sound automatically after a certain amount of time has passed and the fire alarm control unit has not been reset.

Noncoded Alarm

A noncoded system is the simplest type of local alarm. When an alarm-initiating device, such as a smoke detector, sends a signal to the FACP, all of the alarm-signaling devices operate simultaneously **(Figure 2.11)**. The signaling devices usually operate continuously until the FACP is reset. The FACP is not capable of identifying which initiating device triggered the alarm; therefore, building and fire department personnel must walk around the entire facility and visually check to see which device was activated. These systems are only practical in small occupancies with a limited number of rooms and initiating devices.

An FACP serves the premises as a local control unit. This system is found in occupancies that use the alarm signals for other purposes. For instance, schools sometimes use the same bells for class change as are used for fire alarms. The FACP enables the fire alarm to have a sound that is distinct from class bells, thus eliminating confusion as to which type of alarm is sounding. Modern codes do not allow systems such as these; however, older systems that do are still encountered.

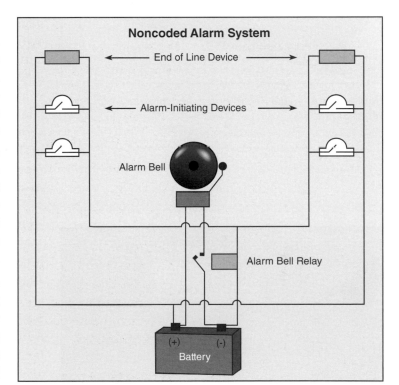

Figure 2.11 A noncoded system cannot differentiate which alarm-initiating device sent the alarm or produce anything other than a continuous signal.

Zoned/Annunciated Alarm

Fire-alarm system annunciation enables emergency responders to identify the general location (zone) of alarm device activation. In this type of system, an annunciator panel, FACP, or a printout visibly indicates the building, floor, fire zone, or other area that coincides with the location of an operating alarm-initiating device **(Figure 2.12, p. 28)**.

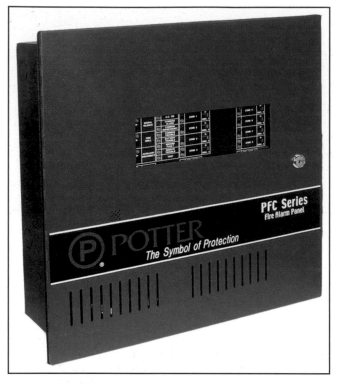

Figure 2.12 A typical FACP for a zoned system. *Courtesy of Potter Signal.*

Alarm-initiating devices in common areas are arranged in circuits or zones. Each zone has its own indicator light or display on the FACP. When an initiating device in a particular zone is triggered, the notification devices are activated, and the corresponding indicator is illuminated on the FACP. This signal gives responders a better idea of where the problem is located.

A remote annunciator panel may be located some distance from the FACP, often in a location designated by the fire department. For example, such an installation may be found at the driveway approach to a large residential retirement complex. This type of annunciator panel usually has a map of the complex coordinated to the zone indicator light.

A zone-indicating system may also be equipped with its own indicator lamp on the FACP, a signal-coding device that is placed into the circuit. This causes the signaling devices to sound in a specific and unique pattern for each zone. This pattern enables employees or fire department personnel to determine the problem zone simply by listening to the audible pattern of the alarms. Usually, the audible pattern is composed of a series of short rings, a brief pause, and a second series of short rings, followed by a long pause. The cycle then repeats. Zoned and annunciated features may be found in conjunction with proprietary fire alarm systems.

Addressable Alarm Systems

Addressable alarm systems display the location of each initiating device on the FACP or annunciator panel **(Figure 2.13)**. This connection ensures that firefighters or building personnel responding to the alarm can pinpoint the specific device that has been activated. Addressable systems reduce the amount of time that it takes to respond to emergency situations. These systems also allow for quick location and correction of malfunctions in the system.

Figure 2.13 Addressable alarm systems identify the specific device that has been activated. *Courtesy of Potter Signal.*

Supervising Station Alarm Systems

A supervising station alarm system is a system that is continuously monitored at a remote location for the purpose of reporting a supervisory, trouble, or alarm signal to the appropriate authorities. Supervising station alarm systems are the predominant types of signal-monitoring systems used in the United States. Types of supervising station systems include auxiliary alarm systems, proprietary systems, central station systems, and remote receiving systems.

Auxiliary Alarm Systems

An **auxiliary alarm system** is connected to a municipal fire alarm system. Alarms are transmitted over this system to a public fire telecommunications center where the appropriate response agencies are selected and dispatched to the alarm **(Figure 2.14)**. Two types of auxiliary fire alarm systems are the local energy system and the shunt-type auxiliary fire alarm system.

A local energy system has its own power source and does not depend on the supply source that powers the entire municipal fire alarm system. In these systems, initiating devices can be activated even when the power supply to the municipal system is interrupted. However, interruption may result in the alarm only being sounded locally and not being transmitted to the fire department telecommunications center. The ability to transmit alarms during power interruptions depends on the design of the municipal system.

Figure 2.14 Auxiliary alarm systems transmit alarms to a public fire telecommunications center where the appropriate fire and emergency services resources are dispatched.

A shunt system is electrically connected to an integral part of the municipal fire alarm system and depends on the municipal system's source of electric power. When a power failure occurs in this type of system, an alarm indication is sent to the fire department communications center. NFPA® 72 allows only manual pull stations and water flow detection devices to be used on shunt systems. Fire detection devices are not permitted on a shunt system.

Auxiliary Alarm System — System that connects the protected property with the fire department alarm communications center by a municipal master fire alarm box or over a dedicated telephone line.

Proprietary Systems

A **proprietary system** is used to protect large commercial and industrial buildings, high-rise structures, and groups of commonly owned facilities such as a college campus or industrial complex in single or multiple locations. Each building or area has its own system that is wired into a common receiving point that is owned and operated by the facility owner. The receiving point must be in a separate structure or a part of a structure that is remote from any hazardous operations **(Figure 2.15, p. 30)**.

The receiving station of a proprietary system is continuously staffed by personnel who are trained in the system's operation and can take necessary actions upon an alarm activation. The operator should be able to automatically summon a fire department response through the use of system controls or may do so manually by using the telephone. Many proprietary systems and receiving points are used to monitor security functions in addition to fire and life safety functions. Modern proprietary systems can be very complex and have a wide range of capabilities, including coded-alarm and trouble signal indications, building utility controls, elevator controls, and fire and smoke damper controls.

Proprietary Alarm System — Fire protection system owned and operated by the property owner.

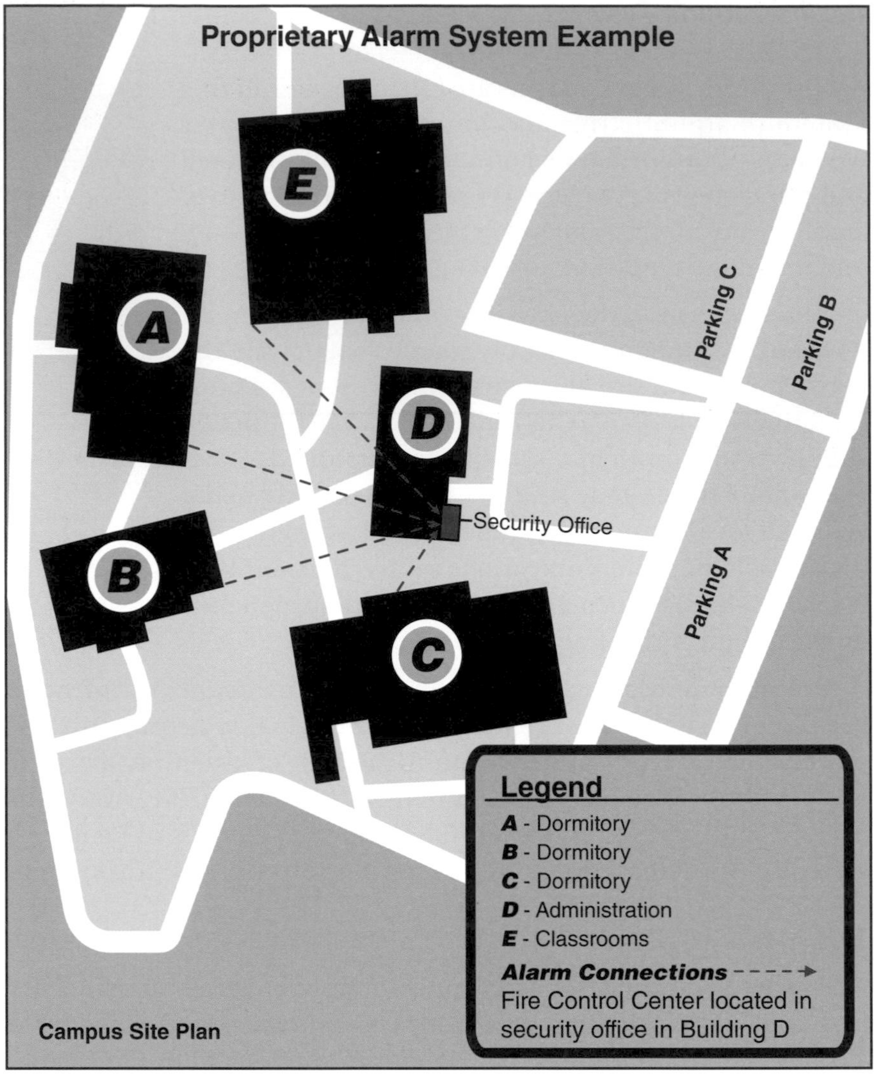
Proprietary Alarm System Example
E
A
D
B
C
Security Office
Parking C
Parking B
Parking A
Legend
A - Dormitory
B - Dormitory
C - Dormitory
D - Administration
E - Classrooms
Alarm Connections
Fire Control Center located in security office in Building D
Campus Site Plan

Figure 2.15 A proprietary alarm system can be used to monitor several sites or locations simultaneously from one receiving point.

Parallel Telephone Systems

A parallel telephone system consists of a dedicated telephone line between each individual alarm box or protected property and the fire department telecommunications center. NFPA® 72 requires that these telephone systems not be used for any other purpose than to relay alarms. These systems are generally not found today due to the existence of private monitoring firms.

Multiplexing systems allow the transmission of multiple signals over a single line. This type of system allows the alarm initiating devices to be identified individually, as in an addressable system, or in a group through the interaction of the fire alarm control panel with each independent device. Remote devices, such as relays, can be controlled over the same line to which initiating and indicating devices are connected. This greatly reduces the amount of circuit wiring needed for large applications.

The control panels for multiplexing systems can range from the simple and relatively inexpensive to the very sophisticated and costly. Some multiplex systems have the added advantage of being able to test the performance of the devices, thereby reducing manpower requirements for preventive maintenance.

Central Station Systems

A **central station system** is monitored by contracted services at a receiving point called the central station. When an alarm is activated at a particular client's location, central station employees receive that information and contact local emergency services and representatives of the occupancy **(Figure 2.16)**. A central station is required to have a runner service. The runner must be a qualified technician who can respond within 2 hours for an alarm or supervisory condition and within 4 hours for a trouble condition. All central station systems and communication methods should meet the requirements set forth in NFPA® 72. Central stations, when meeting the listing requirements, must be listed by an approved listing service.

Remote Receiving Systems

Remote receiving systems are common in jurisdictions that do not require central station systems. These systems are not connected to the emergency services telecommunications center through a municipal alarm box system. Instead, the remote system is connected by another means, usually a telephone line. Where permitted, a radio signal over a dedicated radio frequency may also be used. Remote receiving systems do not incorporate the use of a runner service.

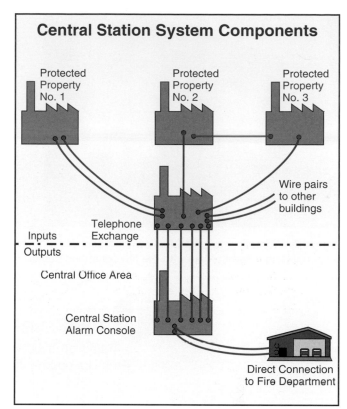

Central Station System Components

Figure 2.16 Components of a central station system.

Depending on local requirements, the fire department may approve other organizations to monitor the remote system. In some small communities, the system is monitored by the local emergency services telecommunications center. This arrangement is particularly common in communities that have volunteer fire departments whose stations are not continuously staffed. In these cases, it is important that emergency services telecommunications personnel are aware of the importance of these alarm signals and trained in the actions that must be taken upon alarm receipt.

Emergency Communications Systems

An *emergency communications system* is a supplementary system that may be placed in a facility in addition to one of the other types of detection and alarm-signaling systems. The purpose of emergency communications systems is to provide a reliable communication system for residents and firefighters. This system may either be a stand-alone system or it may be integrated directly into the overall fire detection and alarm-signaling system. System types include voice notification, two-way communication, and mass notification.

Voice Notification Systems

A *one-way voice notification system* warns building occupants that action is needed and tells them what action to take. This is the type that is most commonly used. Occupants can be directed to move to areas of refuge in the building, leave the building, or stay where they are if they are in an unaffected area.

Central Station System — Alarm system that functions through a constantly attended location (central station) operated by an alarm company. Alarm signals from the protected property are received in the central station and are then retransmitted by trained personnel to the fire department alarm communications center.

Remote Receiving System — System in which alarm signals from the protected premises are transmitted over a leased telephone line to a remote receiving station with a 24-hour staff; usually the municipal fire department's alarm communications center.

Figure 2.17 Emergency phones allow emergency responders to communicate with the fire command center.

Two-Way Communication Systems

A *two-way emergency communications system* allows people at other locations in the building to communicate with the person at the fire command center using either intercom controls or special telephones **(Figure 2.17)**. This system is most helpful to fire-suppression personnel who are operating in a building, particularly in high-rise structures that interfere with portable radio transmissions. Emergency phones are connected in the stairwells and other locations as required by the fire department. These phones enable firefighters to communicate with the incident commander at the fire command center. Most building codes require these systems in high-rise structures.

Other Emergency Communication Systems

In lieu of fire phones, if approved, fire department radio transmissions may be augmented by the installation or use of systems in new or existing buildings. These systems may involve portable or fixed radio repeaters, or a leaky coax. Radio repeaters operate by boosting or relaying fire department radio signals in buildings that may shield or disrupt normal high-frequency radio transmissions due to the weakness of these higher frequencies. Leaky coax systems are similar; however, rather than boosting or relaying radio signals, they simply increase the transmitting capability in these types of buildings by creating a more effective (virtual) antenna that can improve radio communications.

Mass Notification System (MNS) — System that notifies occupants of a dangerous situation and allows for information and instructions to be provided.

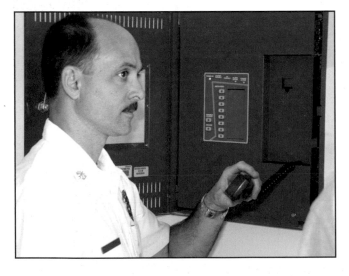

Figure 2.18 Some mass notification systems provide a means for instructions to be given to occupants during an emergency.

Mass Notification Systems

The purpose of a **mass notification system (MNS)** is to provide for emergency communications to a large number of people on a wide-scale basis **(Figure 2.18)**. This communication can be to the occupants of a building, or even an entire community. The events of September 11, 2001, as well as school shootings and other incidents have provided evidence of the need for this type of system.

While the military was the first to implement this technology, today it is being used in public and private facilities. A MNS provides a means to notify a great number of people of an emergency event and allows the ability to provide specific instructions for the appropriate actions to take.

Mass notification systems may be incorporated into an emergency communications system. Those designing this type of system must take into consideration the building being protected, as well as the needs of the occupants. The operation of the building fire alarm must be assigned the highest priority, including the system's own voice message. No other message should override the fire alarm message except under specific circumstances. Specifications for mass notification systems are included in NFPA® 72 and should be consulted for more information.

Manual Alarm-Initiating Devices

Manual alarm-initiating devices, commonly called **manual pull stations** or *pull boxes*, are placed in structures to allow occupants to manually initiate the fire-signaling system. Manual pull stations may be connected to systems that sound local alarms, off-premise alarm signals, or both.

Although manual pull stations come in a variety of shapes and sizes, fire alarm pull stations are required to be red in color with white lettering that specifies what they are and how they are to be used **(Figure 2.19)**. The manual pull station should only be used for fire-signaling purposes unless it is designed for other uses such as communicating with a guard station or activating a fixed fire-suppression system.

According to NFPA® 72, the pull station should be mounted on walls or columns so that the operable part is no less than 3½ feet (1 m) and no more than 4 feet (1.2 m) above the floor so that it can be easily accessed by all occupants. The manual pull station should be positioned so that it is in plain sight and unobstructed. Multistory facilities should have at least one pull station on each floor. In all cases, travel distances to the manual pull station should not exceed 200 feet (60 m).

Figure 2.19 Manual pull stations are required to be red in color and labeled with instructions on their use.

NFPA® 72 also requires that pull stations be placed within 5 feet (1.5 m) of every exit so that facility occupants can activate an alarm while they are exiting the facility (**Figure 2.20**). Most building codes do not require manual alarm-initiating devices in structures that have automatic sprinkler systems and a device that sounds a local alarm when water begins to flow. In this situation, only one pull station is required at an approved location.

Manual pull stations that require the operator to break a small piece of glass with a mallet are no longer recommended. These devices were designed to discourage false alarms and were somewhat effective for that purpose. However, broken glass presents an injury hazard to the operator at a time when an untrained operator is least capable of clear thinking. Polycarbonate covers have taken the place of glass; however, these types of pull stations may still be found in many old structures.

A manual pull station may be protected by an approved wire or plastic basket in areas where it would be subject to damage or accidental activation **(Figure 2.21, p. 34)**. This protective device may be found in gymnasiums, materials handling areas, or in other locations where accidental activation is possible. Some pull stations leave a dye or ultraviolet residue on the activator to discourage malicious false alarms.

Manual pull stations can be single-action or dual-action. Single-action stations operate upon a single motion made by the user. When the station lever is pulled, a lever or other movable part is moved into the alarm position and a corresponding signal is sent to the FACP (**Figure 2.22, p. 34**). A dual-action station requires the operator to perform two steps in order to initiate the alarm. First, the operator must lift a cover or open a door to access the alarm

Figure 2.20 Pull stations are required to be placed within 5 feet (1.5 m) of exits so that an alarm can be signaled during evacuation.

Figure 2.21 Pull stations are often equipped with protective covers to prevent damage or accidental activation. *Courtesy of Potter Signal.*

Figure 2.22 A pull station in the alarm position.

control. Once this action is taken, the alarm level, switch, or button must be operated to send the signal to the FACP. Dual-action manual pull stations may be confusing to certain occupants or operators due to the need to perform two separate steps before an alarm is initiated.

Depending on the type of system they serve, pull stations may be either coded or noncoded. Noncoded pull stations contain an on/off contact switch that is activated when the station is pulled. The pull station sends a continuous signal that sounds until the pull station and system are reset. Coded pull stations send intermittent signals that indicate their locations.

Automatic Alarm-Initiating Devices

Automatic alarm-initiating devices, commonly called *detectors*, continuously monitor the atmosphere of a building, compartment, or area. When certain changes in the atmosphere are detected, such as a rapid rise in heat, the presence of smoke, or a flame signature, a signal is sent to the FACP. These devices are typically very accurate at sensing the presence of the products of combustion they are designed to detect.

Fire protection system designers must take into account the normal activities and environments that occur in any given protected area. Products of combustion may be present when there is no emergency condition. Examples would be a welder's arc in a monitored area or smoke detectors placed in an area where there is dust or excessive moisture **(Figure 2.23)**. These conditions can cause accidental activations of the system.

Four basic types of automatic alarm initiating devices are those that detect heat, smoke, fire gases, and flames. Devices that are a combination of these basic types are also available. The following sections describe these devices.

Fixed-Temperature Heat Detectors

Fire-detection systems using heat-detection devices are among the oldest still in service. **Fixed-temperature heat detectors** are relatively inexpensive compared to other types of systems and are the least prone to false activations. While these devices are reliable, they are typically the slowest to activate under fire conditions. These devices also are not typically resettable and must be replaced after activation.

To be effective, heat detectors must be placed in accordance with their listing. Heat detectors must also be selected at a temperature rating that will give at least a small margin of safety above the normal ceiling temperatures that can be expected in a particular area. According to NFPA® 72, heat-sensing fire detectors must be color-coded and marked with their listed operating temperatures. For more information on detector markings, refer to NFPA® 72.

Because heat is a product of combustion, devices that use one of three primary principles of physics can detect the presence of heat. All heat-detection devices operate on one or more of the three following principles:

1. Heat causes expansion of various materials

2. Heat causes melting of certain materials

3. Heated materials have thermoelectric properties that are detectable

There are a variety of fixed-temperature devices or detectors used in fire detection systems **(Figure 2.24)**. The main types are the fusible link/frangible bulb, bimetallic heat detector, and continuous-line heat detector.

Figure 2.23 A welder's arc can cause accidental activation of the fire detection system.

Fixed-Temperature Heat Detector — Temperature-sensitive device that senses temperature changes and sounds an alarm at a specific point, usually 135°F (57°C) or higher.

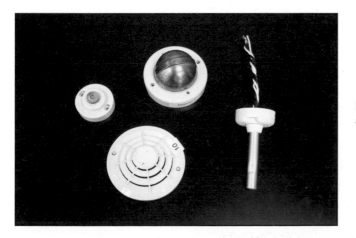

Figure 2.24 A variety of fixed-temperature devices are used in fire detection systems.

Fusible Link/Frangible Bulb

Although **fusible links** and **frangible bulbs** are usually associated with automatic sprinkler systems, they are also used in fire alarm systems. The operating principles of links and bulbs that are used in fire-detection systems are identical to the links and bulbs used with automatic sprinkler systems. It is only their application that differs.

Fusible links are used to hold a spring device in the detector in the open position **(Figure 2.25)**. When the melting point of the fusible link is reached, it melts and drops away, which causes the spring to release and touch an electrical contact that completes a circuit and sends an alarm signal. In order to restore the detector, the fusible link must be replaced.

A frangible bulb is also inserted into a detection device to hold two electrical contacts apart, much like that described for the fusible link. As the temperature increases, the liquid in the bulb expands, compresses the air bubble in the glass, which fractures the bulb that then falls away. The contacts close to complete the circuit and send the alarm. In order to restore the detector, either the frangible bulb or the entire detector must be replaced.

Bimetallic

A **bimetallic** heat detector uses two types of metal that have different heat expansion ratios. Each metal is formed into thin strips and the different metals are then bonded together. One metal expands faster than the other and causes the combined strip to arch when subjected to heat. The amount that it arches depends on the characteristics of the metals, the amount of heat to which they are exposed, and the degree of arch present when in normal position. All of these factors are calculated into the design of the detector.

A bimetallic strip may be positioned with either one or both ends secured in the device. When positioned with both ends secured, a slight bow is placed in the strip. When heated, the expansion causes the bow to snap in the opposite direction **(Figure 2.26)**. Depending on the design of the device, this action either opens or closes a set of electrical contacts that in turn sends a signal to the FACP. Most bimetallic detectors are the automatic resetting type. They need to be checked, however, to ensure that they have not been damaged.

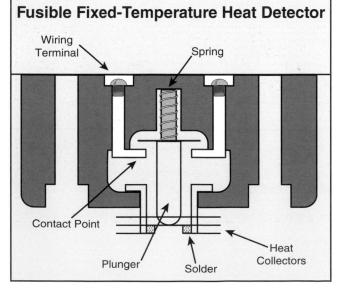

Fusible Fixed-Temperature Heat Detector

Wiring Terminal

Spring

Contact Point

Plunger

Solder

Heat Collectors

Figure 2.25 Cutaway of a fusible link heat detector.

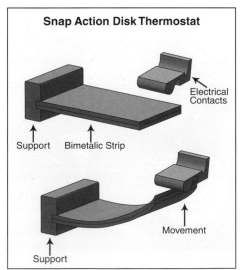

Snap Action Disk Thermostat

Electrical Contacts

Support

Bimetalic Strip

Movement

Support

Figure 2.26 A bimetallic heat detector in the normal and activated positions.

Continuous-Line

Most of the detectors described in this chapter are the spot style; that is, they detect conditions only at the spot where they are located. However, one style of heat-detection device, the *continuous-line (or linear) device*, can be used to detect conditions over a wide area.

There are two models of continuous-line heat detectors available today. One model consists of a conductive metal inner core cable that is sheathed in stainless steel tubing. The inner core and sheath are separated by an electrically insulating semiconductor material, which keeps the core and sheath from touching but allows a small amount of current to flow between them **(Figure 2.27)**. The insulation is designed to lose some of its electrical resistance capabilities at a predetermined temperature anywhere along the line. When the heat at any given point reaches the resistance-reduction point of the insulation, the amount of current transferred between the two components increases. This increase results in an alarm signal being sent to the FACP. This heat-detection device restores itself when the level of heat is reduced.

A second model of continuous-line heat-detection devices uses two wires that are each insulated and bundled within an outer covering **(Figure 2.28)**. When the melting temperature of each wire's insulation is reached, the insulation melts and allows the two wires to touch, which completes the circuit and sends an alarm signal to the FACP. To restore this continuous-line heat detector, the fused portion of the wires must be removed and replaced with new wire.

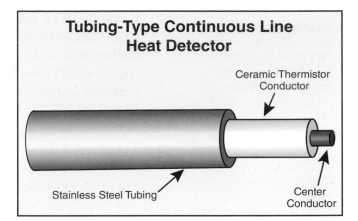

Figure 2.27 One example of a continuous-line heat detector.

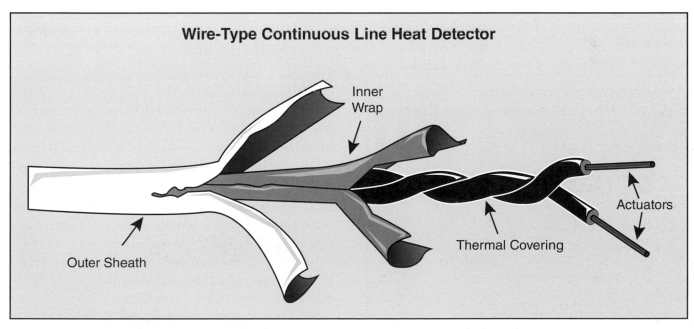

Figure 2.28 Another type of continuous-line heat detector.

Rate-of-Rise Heat Detector

A **rate-of-rise heat detector** operates on the principle that fires rapidly increase the temperature in a given area. These detectors respond in substantially lower temperatures than fixed-temperature detectors. Typically, rate-of-rise heat detectors are designed to send a signal when the rise in temperature exceeds 12°F to 15°F (7°C to 8°C) per minute because temperature changes of this magnitude are not expected under normal, nonfire circumstances.

Most rate-of-rise heat detectors are reliable and typically not subject to false activations. For example, if a rate-of-rise detector is placed near a garage door in an air-conditioned building, an influx of summer air when the door opens will rapidly increase the temperature around the heat detector, causing it to activate. Avoiding such improper placement of a heat detector helps to prevent false activations.

Rate-of-rise heat detectors are designed to automatically reset. Varieties of rate-of-rise heat detectors include the following:

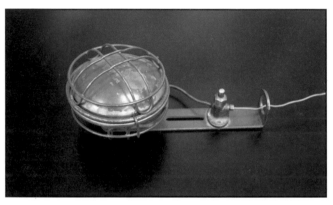

Figure 2.29 Line heat detectors used in pneumatic rate-of-rise systems depend on the change in temperature and increase in pressure of the air in the tubing to activate the alarm system.

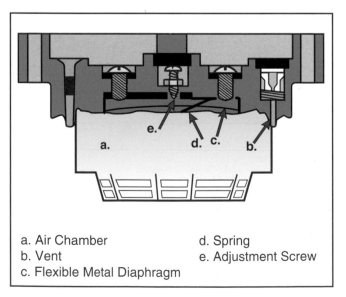

a. Air Chamber
b. Vent
c. Flexible Metal Diaphragm
d. Spring
e. Adjustment Screw

Figure 2.30 The spot heat detector is self-contained in one unit.

- *Pneumatic rate-of-rise line heat detector* — Monitors large areas of a building (**Figure 2.29**). Line heat detectors consist of a system of metal pneumatic tubing arranged over a wide area of coverage. The space inside the tubing acts as a pressurized air chamber that allows the contained air to expand as it heats. These heat detectors contain a flexible diaphragm that responds to the increase in pressure from the tubing. When an area being served by the tubing experiences a temperature increase, the air pressure increases and the heat-detection device operates. The tubing in these systems is limited to about 1,000 ft (300 m) in length. The tubing should be arranged in rows that are not more than 30 ft (9 m) apart and 15 ft (4.5 m) from walls.

- *Pneumatic rate-of-rise spot heat detector* — Operates on the same principle as the pneumatic rate-of-rise line heat detector. The major difference between the two is that the spot heat detector is self-contained in one unit that monitors a specific location (**Figure 2.30**). Alarm wiring extends from the detector back to the FACP.

- *Rate-compensation heat detector* — Contains an outer bimetallic sleeve with a moderate expansion rate. These are designed for use in areas that are subject to regular temperature changes but at rates that are slower than those of fire conditions (**Figure 2.31**). This outer sleeve contains two bowed struts that have a slower expansion rate than the sleeve. The bowed struts have electrical contacts. In the normal position, these contacts do not touch. When the detector is heated rapidly, the outer sleeve expands lengthwise. This expansion reduces the tension on the inner strips and allows the contacts to meet, thus sending an alarm signal to the FACP. If the rate of temperature rise is fairly slow such as 5 to 6°F (2°C to 3°C) per minute, the sleeve expands at a slow rate that maintains tension on the inner strips. This tension prevents unnecessary system activations.

- *Electronic spot-type heat detector* — Consists of one or more thermistors that produce a marked change in electrical resistance when exposed to heat **(Figure 2.32)**. The rate at which thermistors are heated determines the amount of current that is generated. Greater changes in temperature result in larger amounts of current flowing and activation of the alarm system. These heat detectors can be calibrated to operate as rate-of-rise detectors and function at a fixed temperature. Heat detectors of this type are designed to bleed or dissipate small amounts of current, which reduces the chance of a small temperature change activating an alarm.

Figure 2.31 Rate-compensation heat detectors are used in areas that are subject to regular temperature changes.

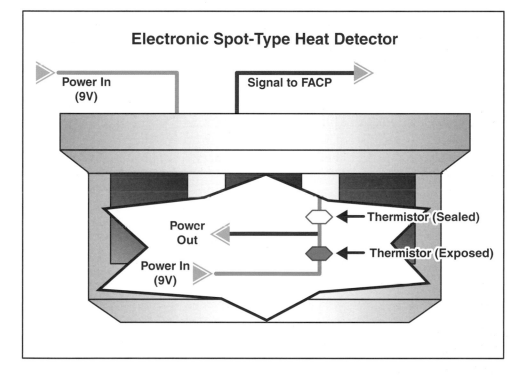

Figure 2.32 The rate at which the temperature of the internal thermistors increases determines the amount of current that is generated to activate and send an alarm signal.

Smoke Detectors

Most people are well aware of the dangers of fire, but less aware of the dangers of smoke inhalation. About 65 percent of fire deaths are attributed to smoke inhalation and not to burns. Smoke and toxic gases spread farther and faster than the heat from flames. When people are asleep, toxic fumes can quickly send victims into a deeper level of unconsciousness. Because of these dangers, an early warning can mean the difference between a safe escape and no escape at all.

From a life-safety standpoint, smoke detectors are the preferred devices in many occupancies such as residences, health care, and institutional care facilities **(Figure 2.33, p. 40)**. This is because smoke detectors sense the presence of products of combustion much more quickly than heat-detection devices.

Figure 2.33 This residential smoke alarm incorporates smoke detection technology as well as a local alarm.

Photoelectric Smoke Detector — Type of smoke detector that uses a small light source, either an incandescent bulb or a light-emitting diode (LED), to detect smoke by shining light through the detector's chamber: smoke particles reflect the light into a light-sensitive device called a photocell.

Many factors affect the performance of smoke detectors such as the type and amount of combustibles, the rate of fire growth, the proximity of the detector to the fire, and ventilation within the area involved.

Smoke detectors are tested and listed based on their performance by a third-party testing service. Regardless of their principle of operation, all smoke detectors are required to respond to the same fire tests. Two basic styles of smoke detectors are in use: photoelectric and ionization. Other smoke detectors such as duct, air-sampling, and video-based are also described in the sections that follow.

Photoelectric

A **photoelectric smoke detector** works satisfactorily on all types of fires and usually responds more quickly to smoldering fires than ionization smoke detectors. Photoelectric smoke detectors are best suited for living rooms and bedrooms because these rooms often contain large pieces of furniture such as sofas, chairs, and mattresses that can burn slowly and create more smoke than flames. Photoelectric smoke detectors automatically reset when conditions return to normal. A photoelectric smoke detector consists of a photoelectric cell coupled with a specific light source. The photoelectric cell functions in one of two ways to detect smoke: projected beam application (obscuration) or refractory application (scattered).

The *projected-beam*, or light obscuration style of photoelectric detector uses a beam of light focused across the area being monitored onto a photoelectric-receiving device such as a photodiode **(Figure 2.34)**. The cell constantly converts the beam into current, which keeps a switch open. When smoke interferes with or obscures the light beam, the amount of current produced is lessened. The detector's circuitry senses the change in current and initiates an alarm when a current change threshold is crossed.

Figure 2.34 Principle of a beam-application photoelectric smoke detector.

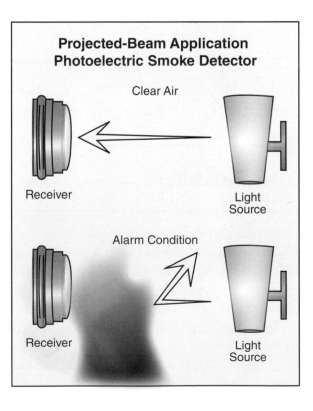

Projected-Beam Application Photoelectric Smoke Detector

Clear Air

Receiver

Light Source

Alarm Condition

Receiver

Light Source

Projected-beam application smoke detectors are particularly useful in buildings where a large area of coverage is desired such as in churches, atriums, or warehouses. Rather than wait for smoke particles to collect at the top of an open area and sound an alarm, the projected-beam application smoke detector is strategically positioned to sound an alarm more quickly. It is important to mount projected-beam application smoke detectors on a stable stationary surface. Any movement due to temperature variations, structural movement, and vibrations can cause the light beams to misalign.

A *refractory application*, or light-scattering smoke detector uses a beam of light from a light-emitting diode (LED) that passes through a small chamber at a point distant from the light source. Normally, the light does not strike the photocell or photodiode. When smoke particles enter the light beam, light strikes the particles and reflects in random directions onto the photosensitive device, causing the detector to generate an alarm signal **(Figure 2.35)**.

Ionization

An **ionization smoke detector** contains a sensing chamber that consists of two electrically charged plates (one positively charged and one negatively charged) and a radioactive source for ionizing the air between the plates **(Figure 2.36)**. A small amount of Americium 241 radioactive material that is adjacent to the opening of the chamber ionizes the air particles as they enter. The ionized particles free electrons from the negative electron plate and the electrons travel to the positive plate. Thus, a small ionization current measurable by electronic circuitry flows between the two plates.

Ionization Detector —
Type of smoke detector that uses a small amount of radioactive material to make the air within a sensing chamber conduct electricity.

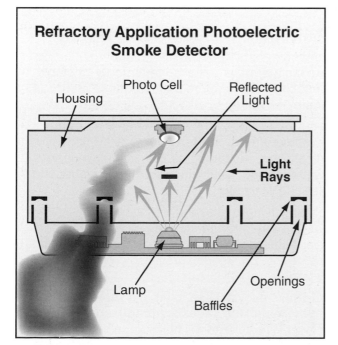

Figure 2.35 Principle of a refractory photoelectric smoke detector.

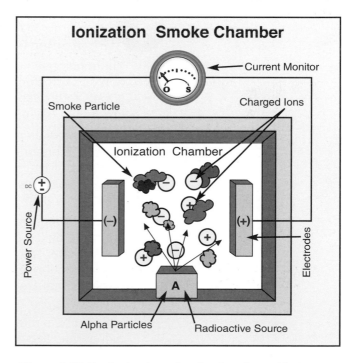

Figure 2.36 Products of combustion interfere with the ionized particles and result in a decrease in the chamber current between the plates, eventually causing an alarm to sound.

Products of combustion, which are much larger than the ionized air molecules, enter the chamber and collide with the ionized air molecules. As the two interact, they combine and the total number of ionized particles is reduced. This action results in a decrease in the chamber current between the plates. An alarm is initiated when a predetermined threshold current is crossed.

Changes in humidity and atmospheric pressure in the room can cause an ionization detector to malfunction and initiate a false alarm. To compensate for the possible effects of humidity and pressure changes, a dual-chamber ionization detector that uses two ionization chambers has been developed and may be found in many jurisdictions. One chamber senses particulate matter, humidity, and atmospheric pressure. The other chamber is a reference chamber that is partially closed to outside air and affected only by humidity and atmospheric pressure. Both chambers are monitored electronically and their outputs are compared. When the humidity or atmospheric pressure changes, both chambers respond equally to the change but remain balanced. When particles of combustion enter the sensing chamber, its current decreases while the reference chamber remains unchanged. The imbalance in current is detected electronically and an alarm is initiated.

An ionization smoke detector works satisfactorily on all types of fires, although it generally responds more quickly to flaming fires than photoelectric smoke detectors. The ionization detector is an automatic resetting type and is best suited for rooms that contain highly combustible materials such as cooking fat/grease, flammable liquids, newspapers, paint, and cleaning solutions.

Photoelectric vs. Ionization: The Continuing Debate

There has been a long-standing debate about which technology (ionization or photoelectric) for smoke alarms is better. The majority of smoke alarms today use ionization technology, and the price for these alarms is lower than that of the photoelectric alarm.

All smoke alarms are tested and listed through nationally recognized testing laboratories and have to meet standards set forth by the industry. The root of the debate over the two technologies lies in their response time. Both alarms have advantages and disadvantages, and these depend on whether the fire is smoldering or flaming.

Smoldering fires are those involving solids, and the rate of progression is limited by the ability of air to penetrate the fuel and increase the heat release rate. The smoke layer from these fires grows slowly and can begin to accumulate well below the ceiling. Smoldering fires also produce smoke particles that are relatively large.

Alternatively, flaming fires have easy access to air and therefore have a higher heat release rate. Smoke and fire gases are produced at a much higher rate, and the smoke layer is hotter and will build at the ceiling faster than a smoldering fire. The smoke particles produced by a flaming fire are relatively small, but they are produced in large amounts.

Studies have shown that an ionization detector responds more easily to the large amount of smoke particles produced by a flaming fire. Photoelectric alarms are more sensitive to the larger particles in smoke found with a smoldering fire. In a 2004 study, the National Institute of Science and Technology (NIST) showed that properly installed and maintained ionization and photoelectric alarms provide enough time to save lives for most of the population under many scenarios. However, it has been shown that photoelectric alarms activate substantially sooner in situations involving smoldering fires.

The 2004 NIST study also compared fire-growth rates of cooking fires and furniture fires to results of a study performed in 1975. While the rate of growth for the cooking fires was relatively the same, the fire-growth rate for furniture fires had been greatly reduced. It was concluded that today's furniture materials and construction were the reasons for the increase in the fire-growth rate. The fire-growth rate directly affects the amount of time occupants have for safe egress.

An important conclusion of the 2004 NIST study was that the available safe-egress time from either smoke alarm would be sufficient if there were more smoke alarms installed in residences to improve audibility and also installed in bedrooms where occupants sleep with their doors closed (See NFPA® 72 for more information on alarm installation). These studies show the importance of fire-safety preparation by highlighting needs for placing smoke alarms in bedrooms, interconnecting alarms, changing alarm tones, and providing better home-escape planning.

In summary, NIST concluded that both ionization and photoelectric alarms provide enough time to save lives; however, ionization alarms may not always sound an alarm during a smoldering fire, even when the room is filled with smoke. On the other hand, ionization alarms will sound earlier in a flaming fire when there is definitely no time to spare.

To improve notification times, manufacturers of smoke alarms have placed both types of technology in a single alarm, thereby providing coverage for both flaming fires and smoldering fires. In addition, NIST is currently conducting research to assess whether modifications need to be made to the standard test method for residential smoke alarms to address the changes in construction and materials of today's furniture products.

Duct

Duct smoke detectors are installed in the return or supply ducts of HVAC systems to prevent smoke and products of combustion from being spread throughout the building. Upon the detection of smoke, the HVAC system will either shut down or transition into a smoke-control mode. The detection of smoke in duct areas is sometimes difficult because the smoke can be diluted by the return air from other spaces or outside air. Duct smoke detectors are no substitute for other types of smoke detectors in open areas.

Air-Sampling

An *air-sampling smoke detector* is a specialty type of smoke detector that is designed to continuously monitor a small amount of air from the protected area. The most common air-sampling detector is the cloud-chamber type (**Figure 2.37**). This detector uses a small air pump to draw sample air into a high-humidity chamber within the detector. The detector then imparts the high humidity to the sample

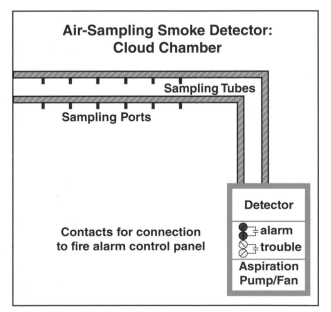

Figure 2.37 The cloud chamber air-sampling smoke detector is the most common type of air-sampling detector.

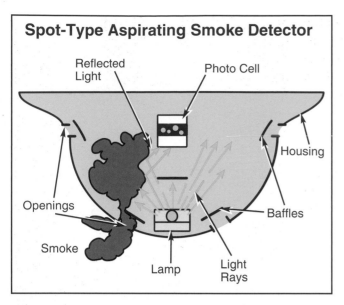

Spot-Type Aspirating Smoke Detector

Reflected Light · Photo Cell · Housing · Baffles · Light Rays · Lamp · Smoke · Openings

Figure 2.38 A similar type of air-sampling smoke detector depends on a photoelectric cell for activation.

and lowers the pressure in the test chamber. Moisture condenses on any smoke particles in the test chamber, which creates a cloud inside the chamber. These detectors can be programmed for several threshold levels as well as wide sensitivity ranges. The detector triggers an alarm signal when the density of this cloud exceeds a predetermined level.

Another type is the continuous air-sampling smoke detector. It is composed of a system of pipes spread over the ceiling of the protected area. A fan in the detector/controller unit draws air from the building through the pipes. The air is then sampled using a photoelectric sensor.

A final type of air sampling detector is the spot-type aspirating smoke detector **(Figure 2.38)**. This type of detector combines the spot-type photoelectric smoke detector with filtered, periodic air sampling. These are designed for use in areas that are very dusty where regular spot-type detectors cannot be used.

Video-Based

Video-based smoke detectors operate on the principle of detecting changes in a digital video image from a camera or a series of cameras. Images are transmitted from a closed-circuit television to a computer that looks for changes in the images. These cameras will work only in a lighted space. They also provide an image to an operator that may be monitoring the system. These systems offer advantages in large, open facilities where there may be a delay in smoke movement and detection.

Flame Detectors

A **flame detector** is sometimes called a *light detector*. There are three basic types of flame detectors:

- Those that detect light in the ultraviolet wave spectrum (UV detectors) **(Figure 2.39)**
- Those that detect light in the infrared wave spectrum (IR detectors)
- Those that detect light in UV and IR waves **(Figure 2.40)**

While these types of detectors are among the fastest to respond to fires, they may be tripped by such nonfire conditions as welding, sunlight, and other bright light sources. Flame detectors must only be placed in areas where these false triggers can be avoided or limited. They must also be positioned so that they have an unobstructed view of the protected area. If flame detectors are blocked, they cannot activate.

To prevent accidental activation from infrared light sources other than fires, an infrared detector requires the flickering action of a flame before it activates to send an alarm. This detector is typically designed to respond to a specific sized fire from a distance determined by the manufacturer.

Flame Detector — Detection and alarm device used in some fire detection systems (generally in high-hazard areas) that detect light/flames in the ultraviolet wave spectrum (UV detectors) or detect light in the infrared wave spectrum (IR detectors).

Figure 2.39 A typical UV flame detector.

There are also video-based flame detectors that work on the same principle as the video-based smoke detectors. The images from the closed-circuit televisions are sent to a computer with software designed to detect the characteristics of a flame. This type of flame detection system may be seen in certain chemical or petroleum facilities.

Fire-Gas Detectors

When a fire ignites in a confined area, it drastically changes the chemical-gas content of the atmosphere in the area. Some of the by-products released by a fire may include the following:

- Water vapor
- Carbon dioxide
- Carbon monoxide
- Hydrogen chloride
- Hydrogen cyanide
- Hydrogen fluoride
- Hydrogen sulfide

Figure 2.40 A combination UV and IR flame detector.

Only water, carbon monoxide, and carbon dioxide are released from all carbonaceous materials that burn. Whether other gases are released depends on the specific chemical makeup of the fuel, so it is only practical to monitor levels of carbon monoxide and carbon dioxide for fire-detection purposes.

Fire-gas detectors implement semiconductors or catalytic elements, which are not used as frequently in other types of detectors. Fire-gas detectors can be found in such places as refineries, chemical plants, and areas of electronic assembly **(Figure 2.41)**.

Fire-Gas Detector —
Device used to detect gases produced by a fire within a confined space.

Combination Detectors

Depending on the design of the system, various combinations of the previously described detection devices are incorporated in a single device. These combinations include fixed-rate/rate-of-rise detectors, heat/smoke detectors, and smoke/fire-gas detectors. *Combination detectors* allow the benefit of both services and increase their responsiveness to fire conditions.

Waterflow Devices

An automatic initiating device that is required on automatic sprinkler systems is the **waterflow device**. This device is designed to activate when water begins to flow through the sprinkler system. Modern designs include electronic flow switches that notify the FACP of a change in the status of the water in the system. This action, in turn, causes the signaling devices to function.

Figure 2.41 Fire-gas detectors are found in industrial occupancies such as refineries and chemical plants.

Waterflow Device —
Detector that recognizes movement of water within the sprinkler or standpipe system. Once movement is noted, the waterflow detector gives a local alarm and/or may transmit the alarm.

Supervisory Devices

Supervisory devices are used to supervise automatic sprinkler systems and monitor the condition of the systems. These devices monitor fire protection control valves that supply the sprinkler and other fire protection systems

Figure 2.42 Supervisory devices are designed to monitor control valves for the fire protection system. This device is used to monitor outside stem and yoke (OS&Y) valves. *Courtesy of Potter Signal.*

(Figure 2.42). If the main water shutoff valve is closed, a supervisory indication is displayed on the FACP. Supervisory devices are also used to monitor air pressure in dry-pipe sprinkler systems, room and water temperature, tank levels, and other devices that may affect the operation of the fire protection system.

Inspection and Testing of Fire Detection and Alarm Systems

To ensure operational readiness and proper performance, fire detection and alarm signaling systems must be tested when they are installed and again on a continuing basis. Tests that are conducted when systems are installed are commonly called **acceptance tests**. Periodic testing is often referred to as **service testing**.

Fire department and fire brigade personnel who routinely conduct inspections need to have a working knowledge of these systems. It is important to keep in mind, however, that these personnel are generally limited to visual inspections and supervision of system tests. They will typically not have to operate or maintain these systems. In most cases, representatives of the company or alarm system contractors actually perform system tests and maintenance. Personnel performing inspection, testing, and maintenance should be qualified and experienced in the types of devices and systems with which they work. Individuals needing more information should consult NFPA® 72.

Acceptance Testing

Acceptance testing is performed soon after the system has been installed or modified to ensure that it meets design criteria and functions properly. The occupant's insurance carrier and/or local codes and ordinances may require acceptance tests. Representatives of the building owner or occupant, the fire department, and the system installer/manufacturer should witness acceptance tests. The fire department representative may be a fire inspector, a staff fire protection engineer, or in some cases the fire marshal.

Some jurisdictions require the system installer or manufacturer to document that the system is ready for inspection by the AHJ. This record prevents the fire inspector or engineer from checking a system that is not ready for acceptance testing.

It is important to test all components of the alarm system. All of the functions of the fire detection and alarm system should be operated and the following tasks performed during acceptance tests:

- *Alarm and trouble modes of system operation* — Check actual wiring and circuitry against the system drawing to ensure that all are connected properly.

- *Fire alarm control panel (FACP)* — Operate all interactive controls at the FACP to ensure that they control the system as designed. Inspect thoroughly to ensure that the FACP is in proper working order **(Figure 2.43)**.

- *Alarm-initiating and signaling devices and circuits* — Check all items for proper operation. Test pull stations, detectors, bells, strobe lights, and other devices to ensure that they are operational. Test each initiating device to ensure that it sends an appropriate signal and causes the system to go into the alarm, supervisory, or trouble mode.

- *Power supplies* — Operate the system on both the primary and secondary power supplies to ensure that both will supply the system adequately.

Restorable heat detectors should be inspected by adhering to approved testing procedures described by the manufacturer. Hair dryers or electric heat guns can be used to effectively and safely test restorable heat detectors. Nonrestorable heat detectors must not be tested during field inspections.

Some combination detectors have both restorable and nonrestorable elements. Caution must be exercised to avoid tripping the nonrestorable element. Nonrestorable pneumatic detectors should be tested mechanically. Those detectors equipped with replaceable fusible links should have the links removed to see whether the contacts touch and send an alarm signal. The links can be replaced following the test.

Figure 2.43 The FACP should be inspected to ensure it is in proper working order.

The manufacturers of smoke, flame, and fire-gas detectors usually have specific instructions for testing their detectors. These instructions must be followed on both the acceptance and service tests. Testing may include the use of smoke-generating devices, aerosol sprays, or magnets. The use of nonapproved testing devices may result in the manufacturer's warranty on the detector being voided.

It is also important to determine the ability of outside entities such as central station, auxiliary, remote station, and proprietary systems to respond to an alarm. The alarm-receiving capability must be verified, and a telecommunication facility must ensure that the signals are properly received.

The results of all tests must be documented to the satisfaction of both the insurance carrier and the fire department. Issuing the alarm system acceptance is typically a preliminary step toward the issuance of a certificate of occupancy. NFPA® 72 contains complete information on acceptance testing.

Service Testing and Periodic Inspection

To ensure that fire detection and alarm systems work when they are needed, the systems must be tested and inspected on a regular basis. Locally adopted fire codes usually mandate that these tests be witnessed by members of the AHJ's inspection division. The actual performance of the tests is the responsibility of the owner/occupant or the fire alarm monitoring company. Periodic tests and inspections are performed on all components of the fire detection and alarm system, including the initiating devices and the FACP.

Detection and Alarm Systems

Fire detection and alarm-signaling equipment should receive a general inspection on a routine basis. The inspections should be conducted by both the AHJ and the building owner/occupant. Because there are many occupancies in a jurisdiction and it is time-consuming to test these systems, it is not always possible for fire department personnel to witness every test.

The building owner/occupant is typically required to test the systems on their own and document the results. At specified intervals, fire department personnel will also be present to witness tests. The procedures addressed in this section should be applied during inspection and testing of the systems. Activities to be performed during an inspection include the following:

- Inspect all wiring for proper support.

- Look for wear, damage, or any other defects that may render the insulation ineffective.

- Inspect conduit for solid connections and proper support wherever circuits are enclosed.

- Check batteries that are used as an emergency power source for clean contact and proper charge.

- Ensure that all equipment, especially initiating and signaling devices, are free of dust, dirt, paint, and other foreign materials. When dust or dirt is found, devices can be cleaned with a vacuum cleaner rather than by wiping. Wiping tends to spread debris around, causing it to settle on electrical contacts. This debris may inhibit the future operation of the system.

- Ensure that access to FACPs, recording instruments, and other devices is not obstructed in any fashion and no objects stored on, in, or around these systems. Many FACPs have storage areas with locking doors for extra relays, lightbulbs, and test equipment.

Alarm-Initiating Devices

Any fire detection and alarm-signaling system will be ineffective unless the alarm-initiating devices are in proper working order and send the appropriate signal to the system control panel. These devices need to be tested per applicable standards to ensure that they are operational. Alarm-initiating device tests can be witnessed by a representative of the AHJ such as an inspector or engineer.

Numerous items need to be checked when testing and inspecting a manual alarm initiating device. The following are recommended tasks when performing an inspection:

- Ensure that each device is installed for its proper application.

- Ensure that access to the device is unobstructed.

- Ensure that each unit is easy to operate (**Figure 2.44**).

Figure 2.44 Manual pull stations should be tested to ensure they are easy to operate.

- Ensure that the unit housing is tightly closed to prevent dust and moisture from entering and disrupting service.

- Remove any chipped, cracked, or otherwise impaired glass and replace.

- Ensure the device's cover or door opens easily and all the components behind the cover or door are in place and ready for service.

- Ensure that components are compatible for use in the system.

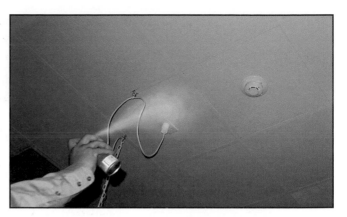

Figure 2.45 Automatic alarm-initiating devices should be tested as required by the AHJ.

Without functional fire detection devices, the most elaborate signaling systems are useless. The reliability of the entire system is, in fact, based largely on the reliability of the detection devices.

Automatic alarm-initiating devices should be checked after installation, after a fire, and at recurring times based on guidelines established by the AHJ or the manufacturer **(Figure 2.45)**. These guidelines are often found in the locally adopted fire code. All detector testing should be in accordance with local guidelines, manufacturer's specifications, and NFPA® 72. Specifically, different manufacturers' components may not be compatible when used in the same system.

Detectors should not be damaged or covered with paint. Detectors found in one or more of the following conditions should be replaced or sent to a recognized testing laboratory for testing:

- Systems that are being restored to service after a period of disuse

- Detectors that are obviously corroded

- Detectors that have been painted, even if attempts were made to clean them

- Mechanically damaged or abused detectors

- Circuits that were subjected to current surges, overvoltages, or lightning strikes

- Detectors subjected to foreign substances that might affect their operation

- Detectors subjected to direct flame, excessive heat, or smoke damage

A permanent record of all detector tests should be maintained by the system owner for at least 5 years. Information that should be included in the record is the test date, the detector type, the location of the detector, the type of test, and the test result.

A nonrestorable fixed-temperature detector cannot be tested periodically. Testing would destroy the detector and require the system to be rendered inoperable until a replacement detector could be located and installed. For this reason, tests are not required until 15 years after the detector has been installed. At that time, 2 percent of the detectors must be removed and laboratory tested. If a failure occurs in one of the detectors, additional detectors must be removed and laboratory-tested. These tests are designed to determine if there is a problem with failure of the product in general or a localized failure involving one or two detectors.

Figure 2.46 Fusible-link detectors are tested by removing the link and observing whether or not the contacts close.

Periodic testing procedures are included in NFPA® 72 and in the fire alarm manufacturer's literature for the system and its components. The following periodic tests are recommended test procedures for the various devices discussed:

- **Restorable heat detection devices** — Test one detector on each signal circuit semiannually. Check as described previously in the section on acceptance testing. A different detector should be selected each time and so noted in the inspection report. Subsequent inspections should include a copy of the previous report to ensure that the same detector is not tested each time.

- **Fusible-link detector with replaceable links** — Check semiannually by removing the link and observing whether or not the contacts close (**Figure 2.46**). After the test, the fusible link must be reinstalled. It is recommended that the links be replaced at 5-year intervals.

- **Pneumatic detector** — Test semiannually with a heating device or a pressure pump. If a pressure pump is used, the manufacturer's instructions must be followed.

- **Smoke detector** — Test semiannually in accordance with the manufacturer's recommendations. The instruments required for performance and sensitivity testing are usually provided by the manufacturer. Sensitivity testing should be performed after the detector's first year of service and every 2 years after that.

- **Flame and gas detection devices** — Require testing by highly trained individuals because they are very complicated devices. Testing is typically performed by professional alarm service technicians on a contract basis.

- **Fire Alarm Control Panels (FACPs)** — Check to ensure that all parts are operating properly. All switches should perform their intended functions and all indicators should illuminate or sound when tested. When individual detectors are triggered, the FACP should indicate the proper location and warning lamps should light. It is important to remember that indicated locations could very well be out of date due to renovations.

Auxiliary devices can also be checked at this time. The auxiliary devices include local evacuation alarms and HVAC functions such as air-handling system shutdown controls, fire dampers, and others. All devices must be restored to proper operation after testing. In connection with these tests, the receiving signals should also be checked. The proper signal and/or number of signals should be received and recorded. Signal impulses should be definite, clear, and evenly spaced to identify each coded signal. There should be no sticking, binding, or other irregularities.

At least one complete round of printed signals should be clearly visible and unobstructed by the receiver at the end of the test. The time stamp should clearly indicate the time of the signal and should not interfere in any way with the recording device.

Timetables

The following list gives a brief synopsis of the inspection and testing requirements for various types of systems and timetable guidelines. If any of these systems use backup electrical generators for emergency power, those generators should run under load monthly for at least 30 minutes.

- *Local alarm systems* — Test in accordance with guidelines established in NFPA® 72 and the manufacturer's recommendations.

- *Central station system* — Test signaling equipment on a monthly basis. Check water-flow indicators, automatic fire detection systems, and supervisory equipment bimonthly. Check manual fire alarm devices, water tank level devices, and other automatic sprinkler system supervisory devices semiannually. When these tests are scheduled, it is important that both facility/building supervisory personnel and central station personnel be notified before the test to prevent them from evacuating occupants or dispatching fire units.

- *Auxiliary fire alarm systems* — Visually inspect and actively test monthly (by the occupant) to ensure that all parts are in working order and the operation of the system results in a signal being sent to the fire department telecommunications center. Test noncoded manual fire alarm boxes semiannually.

- *Remote station and proprietary systems* — Test according to the testing requirements established by the AHJ. Test fire detection components of these systems monthly. Test water-flow indicators semiannually; however, the frequency of testing may depend upon the type of indicator.

- *Emergency communications systems* — Conduct functional tests of the various components in these systems quarterly (by the owner/occupant). Include selected parts of the system that are reflective of what may actually be used during an incident. Check all components at least annually.

Record Keeping

The readiness of fire detection and suppression systems must be documented and records stored systematically. In the performance of an occupation centered on public safety, good record keeping is good business. Records may be written or electronic.

It is crucial for any AHJ to maintain accurate files and records of all occupancies it is responsible for protecting **(Figure 2.47)**. Most often these records are also maintained by the fire alarm contractor and/or the building owner and provide a historical perspective of fire prevention activities within that building or occupancy. Records also provide the basis for all future code compliance and enforcement activities.

Figure 2.47 The AHJ should maintain current and accurate files for all occupancies it is responsible for protecting.

Documents and records that should be maintained include inspection reports, forms, letters, violation notices and citations, plan-review comments, approvals and drawings, fire reports, fire investigations, and compliant permits and certificates issued. It is recommended that records be maintained for the entire life of the building. In other words, if the building is still standing, there should be records kept on that building and these should be readily available in the offices of the local fire agency.

It is very common for a building to house many different owners and occupancies. By maintaining a file on the structure throughout its lifetime, changes that have been made to the structure will be apparent. This practice will assist with evaluation of how any proposed changes may be affected by the previous uses of the structure. Local policy and legal recommendations will dictate how long records should be maintained once the building has been demolished.

Written Records

With the introduction of technological advances, there are more and more agencies that are maintaining records electronically as opposed to written. However, agencies may still have to maintain a large number of written inspection and testing records. This may include older records produced before computerization as well as hard-copy documents.

Each property should have a file that contains copies of all building and inspection records for that property. Each time an inspector has contact with the occupant or the property owner, records of those actions should be added to the file. Files should be kept as current as possible.

Electronic Records

Many fire departments and other inspection agencies are taking advantage of the advances in computer technology to maintain and store testing and inspection records. Computerized record systems can also assist in planning and scheduling future testing and/or inspection needs. The level of sophistication of the computer system should meet the fire department or inspection agency's needs. Most departments use a purchased software program since few departments have the resources to develop their own data management system.

Two primary methods by which data may be logged into the computer system are as follows:

1. Personnel use laptop computers or handheld electronic data-recording equipment while making the inspection and then download the information into the system.

2. Personnel use written forms to record the information while performing the system inspection and then enter the information manually into the computer system upon returning to the office **(Figure 2.48)**.

The ability to electronically record information in the field and then download it into the computer system is the more efficient of the two methods. As further advances in technology make small, portable computer equipment more accessible and affordable, this method will become a more common record-keeping practice.

In determining how to establish the computer management system, the following questions need to be considered:

- How will the information be filed?

- How can the information be retrieved?

- What portion of the information will be stored in a read-only format so that records cannot be accidentally or purposely changed without authorization?

- Which personnel will be given access to retrieve information from the system?
- What information can be released to the public?

Even the best designed computer system is useless unless the individuals using it have received the proper training. It is imperative that those using the data management system be trained on its use so that the information is entered and stored correctly.

Note: Some data management systems link this information directly to emergency dispatch personnel or to the fire apparatus. This information may be used in emergency operations and needs to be as accurate as possible.

Figure 2.48 During inspections, information is often written on paper forms and then entered into the computer system at a later time.

Summary

Whether a fire detection and alarm system is simple or complex, its quick operation is vital if fire emergencies are to be mitigated with as little loss or damage as possible. The earliest systems of ringing bells and crying out warnings have given way to complex electronic systems that alert occupants and emergency services at the same time. Fire-fighting personnel must be familiar with the basic operating features of the detection and alarm systems in their response areas so that they can recognize obvious problems that may be occurring. They also need to know which types of detection systems are appropriate for certain hazards so that the systems that are installed can operate effectively. Because this level of inspection is confined to visual observations and supervision of systems testing, it is also important to involve other trained personnel to ensure system operability and safety.

Review Questions

1. What are the basic components of a fire detection and alarm system?

2. What is an alarm signal, a supervisory signal, and a trouble signal?

3. What are the three basic types of protected premises alarm systems?

4. What are the types of supervising station alarm systems?

5. What is the purpose of an emergency communications system?

6. What are the main types of fixed-temperature heat detection devices?

7. What is the difference between a photoelectric smoke detector and an ionization smoke detector?

8. What byproducts of fire are released by all carbonaceous materials that burn?

9. What is the difference between acceptance testing and periodic or service testing?

10. What are some of the types of documents and records that should be maintained for record keeping?

Smoke Management Systems

Chapter Contents

Divider page photo courtesy of Texas Fire Protection Specialists

chapter 3

Key Terms

FESHE Outcomes

Fire and Emergency Services Higher Education (FESHE) Outcomes: Fire Protection Systems

1. Explain the benefits of fire protection systems in various types of structures.

9. Describe the hazards of smoke and list the four factors that can influence smoke movement in a building.

10. Discuss the appropriate application of fire protection systems.

Smoke Management Systems

Learning Objectives

After reading this chapter, students will be able to:

1. Summarize factors that affect smoke generation and spread.

2. Describe the hazards associated with smoke and other products of combustion.

3. Explain the purpose of a smoke management system.

4. Discuss the history of smoke management systems.

5. Identify the advantages and disadvantages of dedicated smoke control systems and nondedicated smoke control systems.

6. Describe strategies and types of smoke management methods used in buildings.

7. Summarize types of testing for smoke management systems.

8. Discuss smoke management system implications for emergency services personnel.

Chapter 3
Smoke Management Systems

Case History

In November of 1980, a fire broke out in a large hotel and casino facility in Nevada. The fire began in a restaurant inside the facility. While the fire was contained to the lower floors of the structure, smoke was able to quickly spread to the upper floors. The movement of smoke in the structure was due in large part to openings such as elevator shafts and stairwells that did not incorporate standard fire protection practices. In addition, smoke was transported rapidly through the building by the structure's HVAC system. Unfortunately, more than 80 occupants of the structure died. Many of these occupants were on the upper floors, far away from the seat of the fire. The cause of death for the majority of victims was ruled to be smoke inhalation.

Fires can generate enormous amounts of smoke **(Figure 3.1, p. 58)**. Smoke kills the majority of fire victims, and many instances have been reported where the victims were found deceased in areas remote from the fire. In addition, smoke contributes to overall property damage well beyond the seat of the fire.

Contemporary fire protection includes smoke management or smoke control as a part of the overall fire protection system of the building. This is especially true in occupancies such as high-rise buildings, covered malls, buildings with atriums, and warehouses with high-piled storage. **Smoke management** is an all-inclusive term that can include compartmentation, pressurization, exhaustion, dilution, and buoyancy elements. Smoke management also includes smoke barriers and exhaust fans and vents **(Figure 3.2, p. 58)**. **Smoke control** refers to any effort to change the pressure in spaces adjacent to the fire area to compartmentalize or exhaust smoke from the area of the fire's origin.

Presently, smoke management systems are the subject of two different NFPA® standards. NFPA® 92A, *Standard for Smoke-Control Systems Utilizing Barriers and Pressure Differences*, discusses smoke control in stairwells, elevator hoistways, means of egress, and smoke refuge areas. NFPA® 92B, *Standard for Smoke Management Systems in Malls, Atria, and Large Spaces*, examines

Smoke Management System — System that limits the exposure of building occupants to smoke. May include a combination of compartmentation, control of smoke migration from the affected area, and a means of removing smoke to the exterior of the building.

Smoke Control System — Engineered system designed to control smoke by the use of mechanical fans to produce airflows and pressure differences across smoke barriers to limit and direct smoke movement.

Figure 3.1 Smoke generated by fires can often be substantial in quantity and extremely toxic. *Courtesy of the Los Angeles Fire Department – ISTS.*

Figure 3.2 Smoke vents are incorporated into a smoke management system in order to efficiently remove harmful smoke from the structure. *Courtesy of Texas Fire Protection Specialists.*

methods of smoke control in large-volume areas. Additional information on the design of smoke management systems can be found in model codes and American Society of Heating, Refrigerating, and Air-Conditioning Engineers (ASHRAE) Standard 52.

This chapter discusses the principles of smoke movement in buildings and the purpose of smoke management/control systems. The chapter also presents information on the different types of smoke management systems. Testing, maintenance, and inspection of smoke management systems are discussed. Finally, the implications of smoke-control systems for emergency services personnel are addressed from both preincident planning and fireground operations standpoints.

The design of smoke-control systems is a relatively complicated process involving complex mathematical calculations that have been developed from a variety of tests and experiments and computer modeling. This chapter serves as an introduction to the concept of smoke management and control, and the variety of strategies that can be used in the accomplishment of this life-saving technique.

Smoke Movement in Buildings

During a fire, smoke follows the overall air movement within a building. Although a fire may be confined within a fire-resistive compartment, smoke can spread easily to adjacent areas through openings such as construction cracks, pipe penetrations, ducts, open doors, and vertical shafts **(Figure 3.3)**. The heating, ventilating, and air-conditioning (HVAC) system in a building can also provide a path for the communication of smoke through a building.

Smoke generation and spread can be affected by several factors. These factors include the following:

- Stack (chimney) effect
- Buoyancy
- Weather
- Mechanical air-handling systems
- Products of combustion

Some of these factors can cause pressure differences between partitions, compartments, walls, and floors that can result in the spread of smoke. Smoke management systems are designed to take these factors into consideration.

Figure 3.3 Penetrations through the firewall that are not sealed can allow smoke to spread to other areas of the structure.

Stack (Chimney) Effect

The **stack effect** is a naturally occurring vertical movement of air within a building. This phenomenon is also referred to as the *chimney effect*. Stack effect occurs due to a difference in the temperature of a building's interior and the exterior **(Figure 3.4)**. When the outside air is cooler than the building's interior, air will move upward through the building. This occurs in large openings, atriums, stairwells, and vertical shafts. During a fire, smoke movement occurs upward from the fire below through open shafts. Smoke then flows out of the shafts and into the upper floors of a building. This is sometimes referred to as a *normal stack effect*.

A downward flow of air can occur during warmer weather when the air inside is cooler than the air outside. During very hot weather conditions, the stack effect causes air to move vertically downward in buildings. This is referred to as *negative* or *reverse stack effect*. Since stack effect is a function of temperature, it is of greatest concern when the temperature differential between the inside and outside of the building is high.

Buoyancy

Smoke **stratification** and movement occurs in part due to the smoke's buoyancy. This buoyancy is the result of a reduced density and from expansion **(Figure 3.5p. 60)**. The increase in temperature causes the smoke and gases to expand, which causes movement throughout the structure.

In an unsprinklered building, the buoyancy of the fire gases can be a significant factor in the movement of smoke in the building. Smoke forms a layer in the upper part of the fire compartment and spreads vertically out through openings to the floors above. In a sprinklered building, the buoyancy factor is typically reduced due to the cooling of the smoke by the sprinkler.

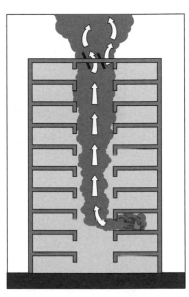

Figure 3.4 The taller the building is, the greater the stack effect.

Stack Effect — Phenomenon of a strong air draft moving from ground level to the roof level of a building. Affected by building height, configuration, and temperature differences between inside and outside air.

Stratification — Formation of smoke into layers as a result of differences in density with respect to height with low density layers on the top and high density layers on the bottom.

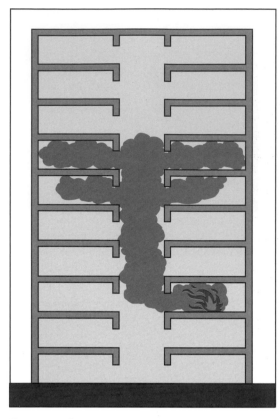

Figure 3.5 Smoke will stratify at a certain level of a structure when its temperature decreases.

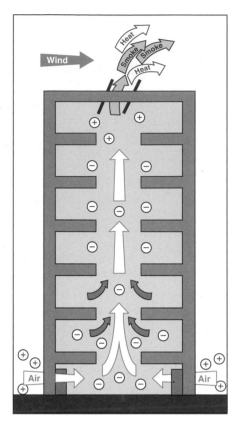

Figure 3.6 Wind moving across the top of a structure can increase stack effect.

Figure 3.7 HVAC systems can contribute to the movement of smoke to other areas of the building.

Weather

Weather can have an effect on smoke movement. While the effect of temperature was discussed previously, wind can have a pronounced effect on smoke movement as well **(Figure 3.6)**. The pressure that wind exerts on the walls of a building depends on the building geometry and wind obstructions. It is possible to calculate wind pressures based on the design of the building.

Buildings that are exposed to the wind experience a positive wind pressure on the wall facing the wind and negative wind pressure on the other three sides. The shape of the building and the velocity of the wind both influence wind pressure. Wind acts to promote horizontal, rather than vertical, air movement through the building. This spreads smoke from the windward side to the leeward side of the building. Of course, in a building with tight exterior walls and no windows, the effects of wind on smoke spread are more likely to be minimal. However, when windows are broken or other large openings are created in exterior walls, the wind will have a greater effect.

Mechanical Air-Handling Systems

HVAC systems can also contribute to smoke movement through a building **(Figure 3.7)**. When a fire starts, an operating HVAC system can transport smoke to every area that the system serves **(Figure 3.8)**. As the fire progresses, the

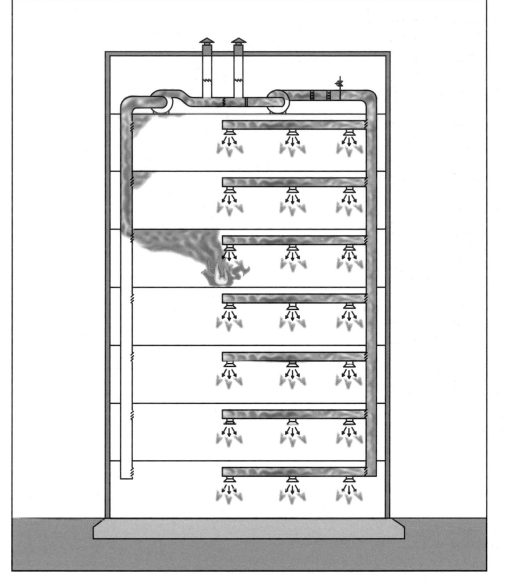

Figure 3.8 The HVAC system can draw the products of combustion into the ducts and transport them throughout the building.

smoke can endanger life, damage property, and inhibit fire-fighting activities. This is especially true with HVAC systems that are not shut down during a fire or that recirculate air through the building. Even if the system is shut down, this alone will not prevent smoke movement through the shafts and return ducts in the system. HVAC ductwork systems are often equipped with duct smoke detectors that will deactivate fans and shut smoke dampers in an effort to limit smoke spread. However, these ductwork systems may cause another avenue for smoke spread from floor to floor.

Products of Combustion

The dangers of smoke and its movement during a fire are related to the products of combustion and the hazards that are created by smoke. Products of combustion can be described very simply as heat, smoke, and sometimes light. However, this description is very deceptive. As fuel burns, its chemical composition is altered, which results in the production of new substances and the release of energy. Structure fires typically involve multiple types of combustible materials and a limited amount of oxygen, which results in incomplete combustion. These factors culminate in the form of complex chemical reactions that produce a wide range of products of combustion. Some of these include toxic and flammable gases, vapors, and particulates.

Heat is a product of combustion that contributes to the spread of fire. This occurs through the preheating of adjacent fuels through **pyrolysis** that makes them more susceptible to ignition. Heat is also a significant cause of injury. The heat from a fire causes burns, damage to the respiratory system, dehydration, and heat exhaustion.

Smoke is an aerosol comprised of fire gases, vapors, and solid particulates. Fire gases such as carbon monoxide (CO) are generally colorless, while vapors and particulates in smoke can give it a variety of colors. Most components of smoke are toxic and present a significant threat to human life. The materials that compose smoke vary from fuel to fuel, but generally all smoke is toxic.

Irritants in smoke are those substances that cause breathing difficulty and inflammation of the eyes, respiratory tract, and skin. Depending on the fuels involved, smoke will contain a wide range of irritating substances.

The toxic effects of smoke inhalation are not the result of any one gas. Instead, these are interrelated effects of all the toxic products present. Three of the more common products of combustion that can be hazardous to building occupants and firefighters include the following:

- *Carbon monoxide (CO)* — This is a byproduct of the incomplete combustion of organic (carbon-containing) materials. CO is probably the most common product of combustion encountered in structure fires. Exposure to CO is frequently identified as the cause of death for civilian fire fatalities and firefighters who have run out of air in their self-contained breathing apparatus (SCBA).

- *Hydrogen cyanide (HCN)* — This gas is produced in the combustion of materials containing nitrogen and is a significant byproduct of the combustion of polyurethane foam (commonly used in furniture and bedding). HCN is also commonly encountered in smoke, although at lower concentrations than CO.

- *Carbon dioxide (CO$_2$)* — This is a product of complete combustion of organic materials and is not toxic in the same manner as CO and HCN. CO$_2$ acts as a simple asphyxiant by displacing oxygen. CO$_2$ also acts as a respiratory stimulant by increasing the respiratory rate.

There are other hazardous components that can be found in smoke depending upon the composition of the fuel that is burning. However, those listed above are the most common.

Hazards of Smoke

There are numerous hazards associated with smoke. As has been mentioned, toxic smoke is responsible for the majority of fire deaths. This is due to the combination of chemicals contained in the smoke. CO causes asphyxiation because it binds much better than oxygen in the blood. Therefore, oxygen is displaced from the hemoglobin in the blood, reducing the amount of oxygen available for use by the body.

Pyrolysis — Thermal or chemical decomposition of fuel (matter) because of heat that generally results in the lowered ignition temperature of the material.

There is also the visual impairment that is caused by dense smoke. The presence of smoke can limit the ability of the building's occupants to evacuate. In addition, smoke presents a more hazardous environment for fire-fighting personnel **(Figure 3.9)**.

Purpose of a Smoke Management System

The purpose of a smoke management system is to reduce occupant deaths and injuries and to aid in the safety of fire-fighters by removing and controlling the spread of smoke. An added benefit of these systems is the reduction of property loss from smoke damage. Most systems are designed for life-safety purposes, though some may be more for property protection, especially where high-value contents are at risk. Smoke management systems are designed to provide a safe escape route, a safe refuge area, or both.

Smoke management systems accomplish their objective in one or more of the following ways:

- Maintaining a tenable environment in the area of egress during the time required for evacuation.
- Controlling and reducing the migration of smoke from the fire area.
- Providing conditions outside the fire zone that will assist emergency response personnel in conducting search-and-rescue operations and in locating and controlling the fire **(Figure 3.10)**.
- Contributing to the protection of life and reduction of property loss.

History of Smoke Management Systems

Historically, the need for smoke management systems dates back to the late 19th and early 20th centuries when large numbers of individuals were killed in several major theater fires. Those incidents included the Brooklyn Theater fire (283 died), the Vienna Ring Theater fire (449 died), and the Theater Royal fire (186 died).

In 1911, a fire broke out in the Palace Theater in Edinburgh. The outcome of this fire was different. Smoke vented through the stage roof, which was credited with the prevention of any loss of life. The buoyancy of the hot smoke forced it through the vent openings. The concept of fan-powered smoke exhaust became the standard for almost all atria in North American buildings.

Figure 3.9 Smoke is a significant hazard to firefighters and SCBA should be worn when operating in areas where smoke is present.

Figure 3.10 Stairwells often provide an area free of smoke for firefighters to stage equipment and prepare to enter the hazardous area.

As was mentioned in Chapter 1, the MGM Grand Hotel fire in Las Vegas, Nevada (November, 1980) is a more modern-day example of the harm that can result from smoke spread. The majority of the 85 deaths that resulted from this incident were in the upper floors of the hotel. In this incident, the smoke from the fire on the first floor spread upward through the HVAC system, the elevator hoistways, and unprotected vertical openings.

In 1973, the Building Department of the City of Atlanta, Georgia, conducted a series of tests of smoke-control systems in the Henry Grady Hotel, a 14-story building that was scheduled to be demolished. The purpose of the tests was to evaluate the effectiveness of various smoke-control systems, including stairwell pressurization (both with and without vestibules) and elevator hoistway pressurization. The stairwell systems were intended to provide a smoke-free egress for building occupants and the elevator system was intended to prevent smoke movement.

The Henry Grady Hotel project demonstrated that pressurization could provide a relatively smoke-free environment for egress given the fire scenarios and systems tested. Additional tests were also performed at the Church Street Office Building in New York City (1973) and a seven-story office building in Hamburg, Germany (1976). In each of these test programs, the smoke management systems were proven to be effective in managing the spread of smoke.

Figure 3.11 Exhaust fans are a component of dedicated smoke-control systems. *Courtesy of Texas Fire Protection Specialists.*

Smoke-Control Strategies

There are a variety of smoke-control methods or strategies. Some are installed as dedicated systems, while others can serve different purposes for the building systems during normal building operations. Different strategies for smoke control include passive (including compartmentation), pressurization, exhaust, opposed airflow, dilution, and zoned.

Dedicated smoke-control systems are those that are intended and specifically listed for smoke-control purposes **(Figure 3.11)**. These systems allow for separate air movement and distribution and do not function under normal operating conditions in the building. Upon activation, these systems operate specifically for smoke control. Advantages of a dedicated smoke-control system include the following:

- Operation and control are generally more simple than other systems.

- Modification of the controls during system maintenance is less likely to occur than with other systems.

- It is less likely to be affected by the modification or failures of other building systems.

While dedicated systems have significant advantages, they may be more costly, require more building space, and have unresolved operational failures since the equipment is not used during normal building operations. Therefore, a strong administrative control and inspection program should be in place.

Nondedicated smoke-control systems are those that share components with other systems such as the building's HVAC system **(Figure 3.12)**. Activation causes the system to change its mode of operation for smoke-control purposes **(Figure 3.13)**. Potential advantages of nondedicated smoke-control systems can include the following:

- Less chance for component failure due to normal use and maintenance

- Lower cost

- Less space needed for mechanical equipment

The disadvantages of the nondedicated system are the elaborate nature of the system control and the possibility of modification of the approved system or smoke controls that might affect the smoke-control function. In addition, all features of the smoke-control system may not be exercised during day-to-day operations.

Figure 3.12 Nondedicated smoke-control systems often share components with the HVAC system to move products of combustion.

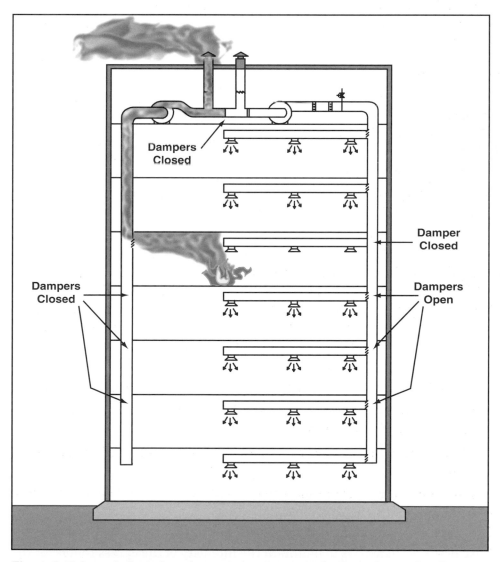

Figure 3.13 A nondedicated smoke-control system operating in the fire mode will exhaust smoke from the fire zone and supply fresh air to adjacent zones.

Passive Systems

Passive smoke control is provided by barriers with sufficient fire endurance to provide protection against fire spread. Walls, partitions, floors, doors, and other barriers provide some level of smoke protection to areas that are a distance from the fire's area of origin **(Figure 3.14)**. Barriers that have sufficient fire rating to be effective have long been used to provide protection against smoke and fire spread. Compartmentation provides passive smoke control between defined areas of a building, which are typically referred to as smoke zones. Codes such as NFPA® 101, *Life Safety Code®*, provide criteria for the construction of smoke barriers, including doors and smoke dampers. The amount of protection against smoke leakage depends on the size and shape of the openings in the barriers and the pressure difference across the barriers.

Examples of passive measures include fire stopping of barrier penetrations, door gasket and drop seals, and stair and elevator vestibules **(Figure 3.15)**. Smoke dampers in HVAC ductwork and automatic door devices may also be used to provide smoke control by compartmentation in a given zone.

Figure 3.14 Fire doors are a common feature of passive smoke control systems.

Pressurization Systems

The pressurization method of smoke control uses mechanical fans and ventilation to create a pressure difference across a barrier such as a wall. Pressure differences across a barrier prevent smoke from infiltrating to the high-pressure side of the barrier. The airflow through the gaps around a door, construction cracks, and other penetrations do not allow smoke to filter from the low-pressure side which is exposed to smoke from the fire.

Pressurization systems must be designed so that they do not create an excessive pressure that may impede safe egress from a building. If the pressure is too great, it may be difficult to open a door. In a sprinklered facility, the pressure required might be less than what is required for an unsprinklered building.

There are two types of pressurization systems: *positive-pressure systems* and *negative-pressure systems*. Positive-pressure systems supply air to the zones adjacent to the zone of the fire's origin. While these systems will adequately contain smoke, they may force smoke from the area of fire origin into unintended areas of the building. In addition, they do not aid in removing smoke from the building.

Negative-pressurization systems serve to exhaust or evacuate the smoke from the area of the fire's origin **(Figure 3.16)**. Shutting down the ventilation system in the areas adjacent to the fire and exhausting smoke from the area of fire origin is the simplest way to accomplish this. With negative-pressure systems, smoke is removed from the building, which improves conditions for both firefighters and occupants.

Figure 3.15 All penetrations of walls or ceilings should be properly sealed. In this photo, the bottom penetration has been sealed with fire stopping, but the upper penetration has not.

Stairwell pressurization is a type of positive pressurization that limits the spread of smoke into stairways in high-rise buildings **(Figure 3.17)**. The purpose of these systems is to maintain the integrity of the egress routes for the building's occupants and provide a smoke-free staging area for firefighters.

These systems are designed so that there is a pressure difference across a closed stairwell door on the fire floor to prevent infiltration of smoke into the stairwell. Stairwell pressurization may be required by building codes or the authority having jurisdiction (AHJ) regardless of any other smoke control methods used in the building.

Pressurization smoke management often depends upon maintaining the integrity of the smoke barriers surrounding the zone to which smoke is to be contained. Maintenance of the integrity of smoke barriers is essential to the operation of the system.

Exhaust Method

The exhaust method is an active smoke-control concept that uses mechanical ventilation along with the properties of smoke to maintain smoke at the highest point in a large space. Because smoke rises to the upper levels of a space, this area serves as a smoke reservoir to contain the smoke **(Figure 3.18, p. 68)**. A properly designed system should allow the smoke to be maintained at a level of 6 to 10 ft (2 m to 3 m) above the highest occupied floor.

The most common method of smoke control by exhaust is to provide for exhaust from the upper portions of the space. NFPA® 92B and ASHRAE Standard 52 provides the calculations necessary to design systems for adequate exhaust of large-volume spaces. In order to have adequate exhaust, air from outside the compartment must be introduced in a sufficient volume to properly exhaust the compartment. These standards also recognize and discuss the use of natural ventilation, although this concept is not widely used in the United States.

The exhaust method is not intended for use in low-ceiling spaces. In addition, sprinkler operation may also adversely impact this approach for spaces with low ceiling heights, because it may cool the smoke causing it to lose buoyancy.

Opposed Airflow Method

For large openings where pressurization is not applicable, a method of smoke control called *opposed airflow* can be used. This method is typically used for openings in a vertical surface, such as doors or hallways, and where the opening is relatively small in comparison to the size of the surface.

With this method, smoke migration from the fire zone is limited by an opposed airflow. High velocity air aimed at the area of fire origin keeps the smoke from migrating into unaffected areas. In order for this method to be successful,

Figure 3.16 Negative pressure fans pull products of combustion from the area of fire origin.

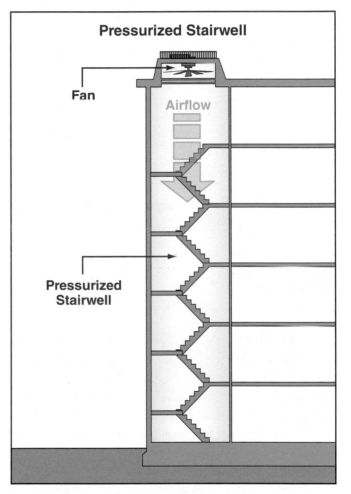

Figure 3.17 A pressurized stairwell incorporates a ventilation system that pushes air into the stairwell. The pressure of this air helps keep smoke out of the stairwell.

Figure 3.18 The exhaust method uses mechanical ventilation to allow smoke to rise to the upper levels of the space. *Courtesy of the McKinney (TX) Fire Department.*

the air has to be of a sufficient velocity and the smoke must be diluted and of a relatively low temperature. It is important that the use of this method not result in airflow toward the fire, which would intensify the fire or interfere with the egress of the building's occupants.

In general, the opposed airflow method does not lend itself to use in buildings. Therefore, caution is recommended for use of the opposed airflow method due to the concern about supplying oxygen to the fire. The successful use of airflow as a smoke management method depends upon the fire being suppressed or the amount of fuel being restricted to limit the size of the fire. These systems have been used successfully in other settings such as subway, railroad, and highway tunnels.

Dilution

Smoke dilution is sometimes referred to as *smoke purging, smoke removal, smoke exhaust*, or *smoke extraction*. This method is most often employed in postfire cleanup but can be used during a fire event. Dilution uses supply and exhaust ventilation to reduce the concentration of smoke within a space. Dilution can be used to maintain tenable conditions in a room that is isolated from the fire by smoke barriers and self-closing doors. When the doors are opened, such as for evacuation or other purposes, smoke will flow through the doorway; however, dilution can improve conditions within the space. This method is not designed to improve conditions within the fire room or areas connected to the fire room.

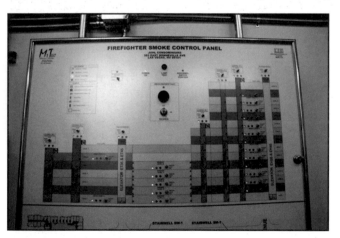

Figure 3.19 Buildings are often divided into zones for smoke control purposes.

During a fire, smoke will flow into areas intended to be protected or safe areas if doors are opened. Supplying outside air into the area can dilute smoke that enters these spaces. This will only be effective in areas remote from the fire. Historically, there has been limited evidence that the use of dilution or smoke purging in the area of the fire will improve conditions in the area or in adjacent spaces.

Zoned Smoke Control

Zoned smoke control is designed to limit the movement of smoke from one compartment of a building to another. With zoned control, a building is divided into a number of smoke zones, separated by partitions and floors **(Figure 3.19)**. During a fire, mechanical fans are used to contain smoke in the zone of fire origin.

In most situations, each floor of a building is chosen as a separate smoke control zone. However, a zone can be multiple floors or can be a division of a single floor. During a fire, all of the nonsmoke zones, or only those adjacent to the fire area, may be pressurized. Changes to building codes have eliminated this requirement and now require that **smokeproof enclosures** be used. However, many zoned systems still exist in older buildings.

Smokeproof Enclosures — Stairways that are designed to limit the penetration of smoke, heat, and toxic gases from a fire on a floor of a building into the stairway and that serve as part of a means of egress.

Zoned smoke control has been used in conjunction with compartmentation. Examples would be hotels or hospitals where the hallways are smoke zones and the rooms are protected by compartmentation.

In some buildings, the HVAC system serves many smoke zones. In order for the system to achieve smoke control, dampers in both the supply and return ducts must be closed as well as the return air damper. For systems where the HVAC system serves only one smoke-control zone, the system has a smoke-zone operational mode and a nonsmoke operational mode.

Firefighters' Smoke-Control Station (FSCS)

A *firefighters' smoke-control station (FSCS)* provides full monitoring and manual control capability over all smoke-control systems and equipment **(Figure 3.20)**. If manual controls are also provided at other building locations for operation of the smoke-control system, the control mode selection from the FSCS should have priority. The smoke-control station should be listed for this purpose.

The purpose of the FSCS is to allow firefighters to have control capability over all smoke-control system equipment or zones within the building. Wherever practical, it is recommended that control be provided by zone, rather than by individual equipment. This will assist firefighters in understanding the operation of the system and will help to avoid problems caused by manually activating equipment in the wrong sequence or by neglecting to control a critical component.

The FSCS should be located in the fire command center or other location as approved by the AHJ **(Figure 3.21)**. Means should be provided to ensure only authorized individuals have access to the FSCS. The FSCS should contain a building diagram that clearly indicates the type and location of all smoke-control equipment as well as the building areas affected by the equipment. The status of the systems and equipment that are activated should be clearly indicated at the FSCS. Manual override switches should be provided at the FSCS to restart or shut down the operation of any smoke-control equipment. More information about the FSCS can be found in Annex D to NFPA® 92A.

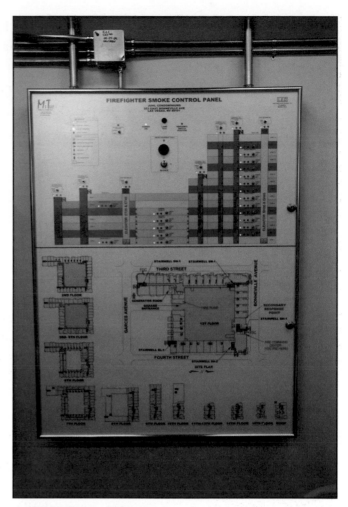

Figure 3.20 A typical FSCS.

Figure 3.21 The FSCS is typically found in the fire command center.

System Testing, Inspection, and Maintenance

Three types of testing that are required to ensure the proper operation of smoke management systems over the life of a building are as follows:

- Acceptance testing
- Periodic performance testing
- Automatic component testing

Figure 3.22 Acceptance testing is done to ensure that the system functions as it was designed. *Courtesy of the McKinney (TX) Fire Department*.

Acceptance Testing

The purpose of acceptance testing is to ensure that the final system installation complies with the specified design and that the system is functioning properly **(Figure 3.22)**. It is recommended that the building owner and building designer share the criteria for smoke control with the local approving agency during the planning stages of the project. These criteria will include a procedure for acceptance testing. Documents should present a clear understanding of the system, its objectives, and the testing procedures. These acceptance testing procedures should be reviewed and approved by the AHJ prior to the final design and acceptance of the system. Prior to the start of the testing process, all building equipment should be in normal operating mode and wind speed and direction information should be recorded.

Acceptance testing of the system should be performed by a qualified inspector as approved by the AHJ. The testing process should include all related fire protection or building systems to the extent that they will affect the operation of the smoke management system. This includes the testing of smoke detection devices and sprinkler water flow switches. The integrity of smoke barriers, partitions, or floor assemblies must be inspected and tested to verify their operational integrity. This may include pressure, air flow, and leakage testing.

Periodic Performance Testing

Periodic testing is critical to ensure proper operation of a smoke management system. During a periodic test, the operational sequence should be verified and performance of the system rechecked. NFPA® 92A and NFPA® 92B recommend annual or semiannual testing, depending on whether the system is nondedicated or dedicated. It is recommended that dedicated systems be tested semiannually and nondedicated systems be tested annually. Persons who are knowledgeable in the operation, testing, and maintenance of the smoke management/control systems should conduct all testing.

Automatic Component Testing

Some systems may be designed or equipped with the ability to perform automatic or self-testing of individual system components. This is conducted through the fire alarm or smoke management panel that controls the system. For example, fans may be automatically turned on just long enough to receive positive confirmation of airflow or dampers may be cycled through an opening and closing process. A report is printed to verify these self-tests, and building personnel should identify any components that fail to operate properly. While this technology is fairly new, it has greatly improved the reliability of smoke management systems in buildings today.

Implications For Emergency Services Personnel

It is important that fire service personnel be familiar with the buildings and systems within their response area. Both prefire planning and fireground operations must consider smoke management systems.

Since the design and installation of smoke management systems should be included in the plans and design of the building, the building office and/or fire department plans examiner should have knowledge of these systems. This information should be shared with emergency responders.

Prefire planning provides responders with specific information about a building's systems and operation. The type of smoke-control methods should be identified, as well as the location of the FSCS and other critical equipment **(Figure 3.23)**. The operation of the system and the implications on fire-suppression operations should be noted by all that might respond to a fire emergency at the location. Firefighters must understand both the capabilities of the particular system and its limitations.

Figure 3.23 The FSCS and other system controls should be identified and located during preincident planning.

The building engineer should be consulted during preincident planning to help firefighters understand how to utilize the system to their best advantage during a fire. This also helps to build a relationship with the engineer, who should be expected to be available for consultation during a fire emergency.

During fire emergencies, responding personnel must understand the overall implications of the smoke management system present in the building. Depending upon the location of the fire, the type of fire, and the location of any occupants, manual adjustments may need to be made to the system's operation. Firefighters need to understand that the building's utilities should not be shut down arbitrarily during a fire because smoke management system equipment may rely on electrical service.

Summary

Smoke control and management is a new concept in fire protection as are the codes that govern its design, installation, and use. These systems provide for a safe means of egress for occupants of a building and a safe environment in which firefighters can function. There are a variety of means or methods of providing for smoke management, and fire department personnel should become familiar with the types of systems employed in their area.

Review Questions

1. What is the stack effect?

2. How can HVAC systems contribute to smoke movement through a building?

3. What are three hazardous products of combustion?

4. What is the purpose of a smoke management system?

5. What are some different strategies for smoke control?

6. What is the difference between dedicated and nondedicated smoke control systems?

7. How do positive pressure smoke control systems work?

8. What is the purpose of a firefighters' smoke control station (FSCS)?

9. What are the three types of testing for smoke control systems?

10. Why should prefire planning and fireground operations consider smoke management systems?

Water Supply Systems

Chapter Contents

Key Terms

FESHE Outcomes

Fire and Emergency Services Higher Education (FESHE) Outcomes: Fire Protection Systems

2. Describe the basic elements of a public water supply system including sources, distribution networks, piping, and hydrants.

3. Explain why water is a commonly used extinguishing agent.

Water Supply Systems

Learning Objectives

After reading this chapter, students will be able to:

1. Explain the extinguishing properties of water.

2. Describe the advantages and disadvantages of using water as an extinguishing agent.

3. Summarize the principles of pressure.

4. Describe different types of pressure as they relate to water.

5. Summarize the principles that govern friction loss in water supply systems.

6. List municipality challenges and responsibilities in providing water.

7. Discuss the basic components of a public water supply system.

8. Discuss private water supply systems.

9. Summarize variables that affect the piping distribution system.

10. Distinguish among valves used on water distribution systems.

11. Discuss types of fire hydrants and their maintenance and testing.

12. Describe how fire hydrants are located and distributed according to the model fire codes.

Chapter 4
Water Supply Systems

Case History

In 2008, a major film studio in California caught fire. While the facility was equipped with sprinklers, the system was quickly overwhelmed due to the amount of fire and the fuel load in the structures. Firefighters on scene attempted to extinguish the fire but were unable to do so because of extremely low water pressure. In some instances, firefighters were unable to even reach the fire with their water streams. While the water supply came from the municipal provider, the film studio owned and maintained its own water distribution system inside the facility. Efforts by the municipal water provider to increase water pressure had little effect. Eventually, fire department personnel had to bring in water by tankers and draft from adjacent water sources in order to obtain enough water to bring the fire under control.

Despite the development of many new innovations and techniques for controlling and fighting fires, water continues to be the primary extinguishing agent. This is due to its availability, affordability, and effectiveness. In addition, water is easily stored and can be conveyed long distances. Because water continues to be the primary extinguishing agent, it is important that the process of obtaining and moving water be understood by fire service personnel.

Basic knowledge about water and water supply systems is necessary when individuals design, inspect, and maintain fire protection systems. The quantity of water necessary and the corresponding pressures must be determined in order for the fire protection to be adequate for a building or facility. Individuals must examine existing water supply systems to determine whether the calculated needs can be met or alternative water supplies are necessary.

It is not possible to cover the specifics of water supply systems for each individual community. It is the responsibility of those who are designing, installing, inspecting, or maintaining fire protection systems to determine the adequacy of local water supplies and systems. This chapter provides information on water supply systems in communities and fire and emergency services operations in the use and support of those systems.

Water as an Extinguishing Agent

Water is a compound of hydrogen and oxygen formed when two parts of hydrogen (H) combine with one part of oxygen (O) to form H_2O. Water exists in a liquid state between 32°F and 212°F (0°C and 100°C). Below 32°F (0°C), water experiences a phase change to a solid state of matter called *ice*. Above 212°F (100°C), it vaporizes into water vapor or steam **(Figure 4.1)**. Water cannot be seen in vapor form; it only becomes visible as it rises away from the surface of the liquid water and begins to condense.

Water is considered to be incompressible and its weight varies at different temperatures. Water's *density*, or its weight per unit of volume, is measured in pounds per cubic foot (lb/ft³) (kg/m³) or pounds per gallon (lb/gal) (kg/L). Water is heaviest and has its highest density close to its freezing point. Water is lightest and has its lowest density close to its boiling point. For fire protection purposes, ordinary fresh water is generally considered to weigh 62.4 lb/ft³ (1 000 kg/m³) or 8.34 pounds per gallon (1 kg/L).

Water is widely used in the fire service as an extinguishing agent because it is plentiful and has unique properties that help in extinguishment. The following sections will discuss these properties.

Extinguishing Properties of Water

Water has the ability to extinguish fire in several ways. The primary way is by cooling, namely absorbing heat from the fire. Another way is smothering, which occurs when oxygen is excluded from the fire. Smothering works especially well on the surface of burning liquids that are heavier than water. A smothering effect also occurs to some extent when water converts to steam in a confined space.

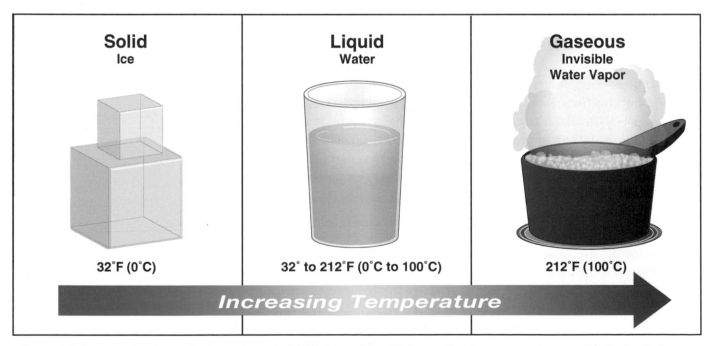

Solid Ice	Liquid Water	Gaseous Invisible Water Vapor
32°F (0°C)	32° to 212°F (0°C to 100°C)	212°F (100°C)

Increasing Temperature

Figure 4.1 An example of water in three states: solid, liquid, and gas. The transition from one state to another is due to the increase or decrease in the temperature of the water.

The extinguishing property of water is affected by the following four factors:

- Law of Specific Heat
- Law of Latent Heat of Vaporization
- Surface area of the water
- Specific gravity

Law of Specific Heat

Water is not only noncombustible, but it can also absorb large amounts of heat. **Specific heat** is a measure of the heat-absorbing capacity of a substance, which means the amount of heat required to change a unit quantity of a substance by 1 degree in temperature. The amount of heat transfer is usually measured in **British thermal units (Btu)** or in the International System of Units (SI) as joules (J). A Btu is the amount of heat required to raise the temperature of 1 lb of water 1°F. The joule, also a unit of work, has taken the place of the calorie in the SI heat measurement.

The specific heat of a substance is the ratio between the amount of heat needed to raise the temperature of a specified quantity of a material and the amount of heat needed to raise the temperature of an identical quantity of water by the same number of degrees. The specific heat of different substances varies. **Table 4.1** shows various fire-extinguishing agents and the specific heat comparisons of each with water. These comparisons are based upon the weight of the agent. Dividing the specific heat of water by the specific heat of another extinguishing agent demonstrates that water is clearly the best material for absorbing heat. For example, water absorbs five times as much heat as does an equal amount of carbon dioxide.

> **Specific Heat** — The amount of heat required to raise the temperature of a specified quantity of a material and the amount of heat necessary to raise the temperature of an identical amount of water by the same number of degrees.

> **British Thermal Unit (Btu)** — Amount of heat energy required to raise the temperature of 1 lb of water 1°F. *1 Btu = 1.055 kilo joules (kJ).*

Table 4.1
Specific Heat of Extinguishing Agents

Agent	Specific Heat
Water	1.00
Calcium chloride solution	0.70
Carbon dioxide (solid)	0.12
Carbon dioxide (gas)	0.19
Sodium bicarbonate	0.22

Law of Latent Heat of Vaporization

The **latent heat of vaporization** is the quantity of heat absorbed by a substance when it changes from a liquid to a vapor. The temperature at which a liquid absorbs enough heat to change to vapor is known as the *boiling point*. At sea level, water begins to boil or vaporize at 212°F (100°C). Vaporization, however, does not completely occur the instant water reaches the boiling point. Each lb of water requires approximately 970 Btu of additional heat to completely convert into steam.

> **Latent Heat of Vaporization** — Quantity of heat absorbed by a substance at the point at which it changes from a liquid to a vapor.

The latent heat of vaporization is significant in fire fighting because the temperature of the water is not increased beyond 212°F during the absorption of the 970 Btu for every lb of water. For example, a gallon of water weighs 8.34 lbs. At 60°F, it requires 152 Btu to raise the temperature of each pound to 212°F (212 – 60 = 152). Therefore, 1 gal of water absorbs 1,268 Btu to get to 212°F (152 × 8.34). Because the conversion to steam requires another 970 Btu/lb, an additional 8,080 Btu will be absorbed through this process. This means that 1 gal of water will absorb 9,346 Btu of heat if all the water is converted to steam **(Figure 4.2, p. 80)**. In fire-fighting terms, if water from

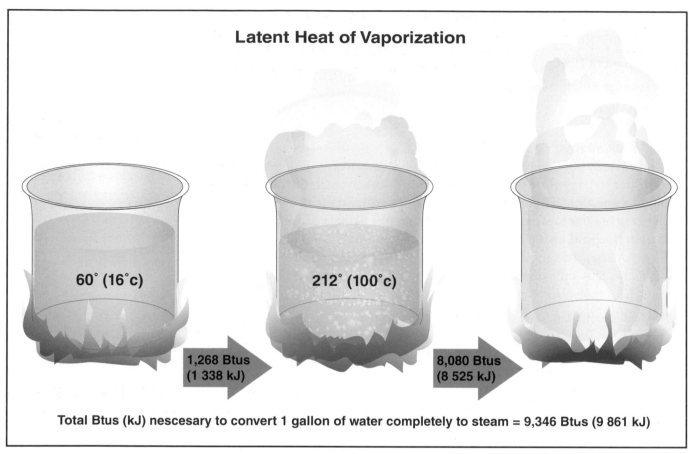

Latent Heat of Vaporization

60° (16°c)

212° (100°c)

**1,268 Btus
(1 338 kJ)**

**8,080 Btus
(8 525 kJ)**

Total Btus (kJ) nescesary to convert 1 gallon of water completely to steam = 9,346 Btus (9 861 kJ)

Figure 4.2 An example of the latent heat of vaporization.

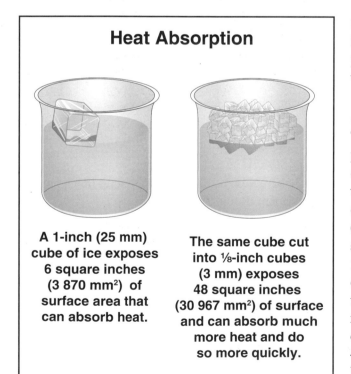

Heat Absorption

**A 1-inch (25 mm)
cube of ice exposes
6 square inches
(3 870 mm²) of
surface area that
can absorb heat.**

**The same cube cut
into ⅛-inch cubes
(3 mm) exposes
48 square inches
(30 967 mm²) of surface
and can absorb much
more heat and do
so more quickly.**

Figure 4.3 Even though the same amount of ice is used, more heat can be absorbed by the smaller cubes due to their increased surface area over the larger cube.

a 100-gallon-per-minute (gpm) (400 L/min) fog nozzle is projected into a highly heated area, it can absorb approximately 934,600 Btu of heat per minute if all of the water is converted to steam.

Surface Area of Water

The speed with which water absorbs heat increases in proportion to the water surface exposed to the heat. For example, if a 1-in cube of ice is dropped into a glass of water, it will take a considerable amount of time for the ice cube to absorb its capacity of heat, or melt; because only 6 square inches (in^2) of ice are exposed to the water. If the same cube of ice is divided into smaller pieces, such as 1/8-in cubes, and these are dropped into the water, 48 in^2 of the ice are exposed. Although the smaller cubes equal the same mass of ice as the larger cube, the smaller cubes melt faster. The larger overall surface area of crushed ice causes it to melt in a drink faster than cubed ice (**Figure 4.3**). The same theory applies to water in its liquid state. If water is divided into many drops, the rate of heat absorption increases hundreds of times.

The expansion capability of water as it converts to steam is another characteristic that aids in fire fighting. Expansion helps cool the fire by driving heat and smoke from the area. The temperature of the fire area will determine the amount of heat expansion. At 212°F (100°C), water expands approximately 1,700 times its original volume.

Specific Gravity

The density of liquids in relation to water is known as **specific gravity**. Water is given a value of 1. Liquids with a specific gravity less than 1 are lighter than water and therefore float on water. Those with a specific gravity greater than 1 are heavier than water and will sink to the bottom. If the other liquid also has a specific gravity of 1, it mixes evenly with water. Most flammable liquids have a specific gravity of less than 1 **(Table 4.2)**.

Water can also smother a fire when it floats on liquids that have a higher specific gravity than water, such as carbon disulfide. If the material is water soluble, such as alcohol, the smothering action is not likely to be effective. Water may also smother fire by forming an emulsion over the surface of certain combustible liquids. When a spray of water agitates the surface of these liquids, the agitation causes the water to be temporarily suspended in emulsion bubbles on the surface. These bubbles then smother the fire. The emulsifying ability of a liquid is dependent upon other properties it possesses. Therefore, if a firefighter directs water on a flammable liquid fire improperly, the whole fire can just float away on the water and ignite everything in its path. The use of foam can control this situation because it floats on the surface of the flammable liquid and smothers the fire.

Specific Gravity — Weight of a substance compared to the weight of an equal volume of water at a given temperature. A specific gravity less than 1 indicates a substance lighter than water; a specific gravity greater than 1 indicates a substance heavier than water.

Advantages and Disadvantages of Water

Some advantages of using water as an extinguishing agent include the following:

- *Heat-absorbing capacity* — Water has a greater heat-absorbing capacity than other common extinguishing agents due to the relatively large amount of heat required to change water to steam.

- *Surface area* — Water applied to a fire by fog patterns and deflected solid streams greatly increases its surface area and causes heat to be absorbed more rapidly.

- *Availability* — In most areas, water is plentiful and readily available.

- *Cost* — Water is relatively inexpensive when compared to other commercially available extinguishing agents.

Disadvantages of using water as a fire-extinguishing agent include the following:

- *High surface tension* — Water does not readily soak into dense materials. The addition of wetting agents to water can decrease this surface tension and improve penetration.

Table 4.2
Specific Gravity of Common Substances

Substance	Specific Gravity at 68°F
Water	0.998
Commercial Solvent	0.717
Carbon Tetrachloride	1.582
Medium Lubrication Oil	0.891
Medium Fuel Oil	0.854
Heavy Fuel Oil	0.908
Regular Gasoline	0.724
Turpentine	0.862
Ethyl Fuel	0.789
Benzene	0.879
Glycerin	1.262
Light Machinery Oil	0.907
Air	0.0012
Ammonia	0.0007
Carbon Dioxide	0.0018
Methane	0.0017

- *Reactivity* — Water is reactive with certain substances such as combustible metals (**Figure 4.4**).

- *Low opacity and reflectivity* — Water allows radiant heat to easily pass through it.

- *Freezing point* — The freezing of water can pose a danger to suppression systems. In cold climates, ice may form in and on equipment, causing the potential for malfunction (**Figure 4.5**).

- *Conductivity* — Water conducts electricity, which can be a hazard when fire fighting operations are taking place around energized electrical equipment.

- *Contaminated runoff* — Water can carry contaminants away from the fire scene.

- *Weight* — Excessive water weight can contribute to structural instability.

Figure 4.4 Placards in some storage facilities indicate that no water should be used due to its reactivity with the item being stored. *Courtesy of Rich Mahaney.*

Figure 4.5 In cold climates, ice often forms on equipment during fire fighting operations. This ice accumulation leads to the instability of equipment and makes the scene hazardous for personnel. *Courtesy of the Chicago (IL) Fire Department.*

Water Pressure

Pressure — Force per unit area exerted by a liquid or gas measured in pounds per square inch (psi) or kilopascals (kPa).

In fire service terminology, **pressure** refers to the force that moves water through a conduit — either a pipe or a hose. Pressure is defined as force per unit area in a liquid or gas. It can be expressed in pounds per square inch (psi), pounds per square foot (psf), or kilopascals (kPa). In order to determine pressure, it is necessary to know the weight of water and the height that a column of water occupies. The weight of 1 cubic foot (ft^3) of water is approximately 62.4 lb. Because 1 square foot (ft^2) contains 144 in^2, the weight of water in a 1 in^2 column of water 1 ft high equals 62.4 lb divided by 144 in^2, or 0.433 lb. Therefore, a 1 in^2 column of water 1 ft high exerts a pressure 0.433 psi at its base (**Figure 4.6**).

In metrics, a cube that is 0.1 m × 0.1 m × 0.1 m (a cubic decimeter) holds 1 L of water. The weight of 1 L of water is 1 kg. The cube of water exerts 1 kPa (1 kg) of pressure at the bottom of the cube. One cubic meter of water holds 1 000 L of water and weights 1 000 kg. Because the cubic meter of water is comprised of 100 columns of water, each 10 decimeters tall, each column exerts 10 kPa at its base.

There are several principles that determine how fluids act when pressure is applied. These principles are introduced in the following sections.

Principles of Pressure

There are six basic principles that determine the action of pressure upon fluids. It is very important that fire and emergency services personnel clearly understand the various sources of pressure. These principles are as follows:

- *First Principle: Fluid pressure is perpendicular to any surface on which it acts* — In a flat-sided vessel containing water, the pressure exerted by the weight of the water is perpendicular to the walls of the container **(Figure 4.7)**. If this pressure is exerted in any other direction, the water would start moving downward along the sides and rising in the center.

- *Second Principle: Fluid pressure at a point in a fluid at rest is the same intensity in all directions* — In other words, fluid pressure at a point in a fluid at rest has no direction.

- *Third Principle: Pressure applied to a confined fluid from without is transmitted equally in all directions* — This principle is illustrated by viewing a hollow sphere to which a water pump is attached **(Figure 4.8, p. 84)**. A series of gauges is set into the sphere around its circumference. When the sphere is filled with water and pressure is applied by the pump, all gauges will register the same pressure. This is true if they are on the same grade line with no change in elevation.

- *Fourth Principle: The pressure of a liquid in an open vessel is proportional to its depth* — This principle is illustrated by observing three vertical containers, each 1 in² with different depths of water. If the first container has 1 ft (300 mm) of water, the second 2 ft (600 mm) and the third 3 ft (900 mm), the pressure at the bottom of the second container will be twice that of the

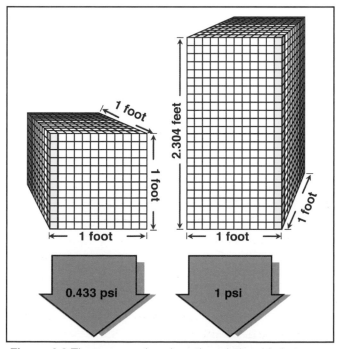

Figure 4.6 These examples show the relationship between height and pressure.

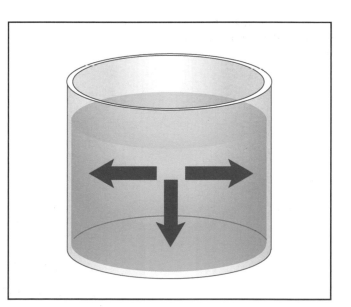

Figure 4.7 Fluid pressure is perpendicular to any surface on which it acts.

first, and the pressure at the bottom of the third will be three times that of the first **(Figure 4.9)**.

- *Fifth Principle: The pressure of a liquid in an open vessel is proportional to the density of the liquid* — If one container holds mercury and the other holds water, it would take 13.55 in (344 mm) of water to produce the same pressure as 1 in (25 mm) of mercury. Mercury is therefore much denser than water **(Figure 4.10)**.

- *Sixth Principle: The pressure of a liquid on the bottom of a vessel is independent of the shape of the vessel* — Regardless of the shape of the container and its opening, the pressure of the water column is the same for a given height.

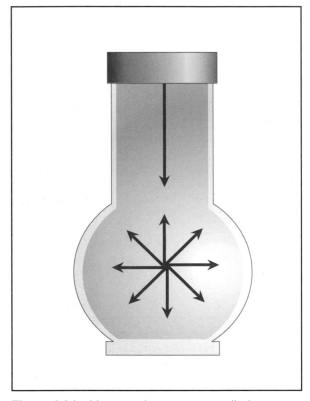

Figure 4.8 In this example, pressure applied to a confined fluid will be transmitted equally in all directions.

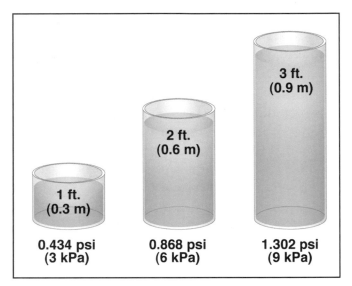

Figure 4.9 The pressure of a liquid in an open vessel is proportional to its depth. For example, the pressure in the second vessel is twice the pressure in the first. The pressure in the third vessel is three times the pressure in the first.

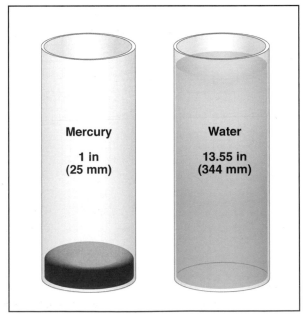

Figure 4.10 In this example, one inch of mercury has the same pressure as 13.55 inches of water because mercury has a much greater density.

Types of Pressure

A variety of terms are used for different types of pressure that are encountered in water supply systems. It is important to properly identify these terms so that they can be understood when working with water supply systems. The types of pressure include the following:

- Atmospheric pressure
- Head pressure
- Static pressure
- Normal operating pressure
- Residual pressure
- Flow pressure (Velocity Pressure)

Atmospheric Pressure

The pressure that is exerted on the earth by the atmosphere itself is called **atmospheric pressure**. The atmosphere surrounding the earth has depth and density and exerts pressure upon everything. Atmospheric pressure is greatest at low altitudes. At sea level, atmospheric pressure is 14.7 psi (101 kPa). This is considered standard atmospheric pressure. As the altitude increases, atmospheric pressure decreases **(Figure 4.11)**. The readings on most pressure gauges indicate the psi (kPa) above the existing atmospheric pressure. For example, a gauge reading of 10 psi (69 kPa) at sea level is actually measuring 24.7 psi (170 kPa). Engineers distinguish between a gauge reading and total atmospheric pressure by writing *psig*, which is the gauge reading. A notation of *psia* means the absolute or the total atmospheric pressure. Any pressure less than the current atmospheric pressure is a vacuum.

Atmospheric Pressure — Force exerted by the atmosphere at the surface of the earth due to the weight of air. Atmospheric pressure at sea level is about 14.7 psi (101 kPa). Atmospheric pressure increases as elevation is decreased below sea level and decreases as elevation increases above sea level.

Head Pressure

The term **head** in the fire service is another way of expressing pressure. It refers to the height that a pressure can lift a column of liquid. If a water supply is 100 ft (30 m) above a hydrant discharge opening, it is said to have 100 ft (30 m) of

Head — Alternate term for pressure, especially pressure due to elevation. For every 1-foot increase in elevation, 0.434 psi is gained (for every 1-meter increase in elevation, 9.82 kPa is gained). Also called Head Pressure.

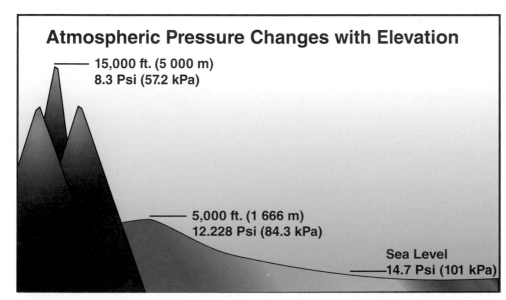

Atmospheric Pressure Changes with Elevation

— 15,000 ft. (5 000 m)
8.3 Psi (57.2 kPa)

— 5,000 ft. (1 666 m)
12.228 Psi (84.3 kPa)

Sea Level
—14.7 Psi (101 kPa)

Figure 4.11 As altitude increases, atmospheric pressure decreases.

head. To convert this to head pressure, divide the number of feet by 2.304. The result is 43.4 psi at the hydrant **(Figure 4.12)**. In metrics, divide the number of meters by 0.1 to get head pressure in kPa.

In the fire service, elevation affects the pressure or flow of water. Elevation refers to the centerline of the pump or the bottom of a static water supply source above or below ground level. The height of a water supply above the discharge orifice is called the **elevation head**. If the nozzle or orifice flowing water is below the pump, there will be a pressure gain. Alternatively, if the nozzle or orifice is above the pump, then there will be a pressure loss. This is referred to as *elevation head pressure*.

Static Pressure

Static pressure is defined as stored potential energy available to force water through pipes, fittings, fire hose, and adapters. The term *static* means at rest or without motion. Pressure on water may be produced by an elevated water supply, by atmospheric pressure, or by a pump. If the water is not moving, the pressure exerted is static. True static pressure is rarely found in a municipal water system because water is always flowing due to customer demand. However, the pressure in a water system before water flows from a hydrant is considered static pressure.

Normal Operating Pressure

Normal operating pressure is that pressure found in a water distribution system during normal consumption demands. As soon as water starts to flow through a distribution system, static pressure no longer exists. The demand for water consumption fluctuates continuously, causing water flow to increase or decrease in the system.

Residual Pressure

Residual pressure is that part of the total available pressure not used to overcome friction loss or gravity while forcing water through pipes, fittings, fire hose, and adapters. Residual means a remainder or that which is left. In a water distribution system, residual pressure varies according to the amount of water flowing, the size of the pipe, and any other restrictions that are present. For example, during a hydrant fire flow test, residual pressure represents the pressure left in a distribution system within the vicinity of one or more flowing hydrants. In a water distribution system, residual pressure varies according to the amount of water flowing from one or more hydrants, water consumption demands, and the size of the pipe. It is important to remember that residual pressure is measured at the location where a pressure reading is taken, not at the point of water flow.

Flow Pressure (Velocity Pressure)

Flow pressure is the forward velocity pressure while water is flowing. The forward velocity or flow pressure can be measured by using a pitot tube and gauge **(Figure 4.13)**. If the size of the opening is known, a firefighter can use the measurement of flow pressure to calculate the quantity of water flowing in gpm or L/min.

Elevation Head — The height of a water supply above the discharge orifice.

Static Pressure — (1) Potential energy that is available to force water through pipes and fittings, fire hose, and adapters. (2) Pressure at a given point in a water system when no water is flowing.

Residual Pressure — Pressure at the test hydrant while water is flowing. It represents the pressure remaining in the water supply system while the test water is flowing and is that part of the total pressure that is not used to overcome friction or gravity while forcing water through fire hose, pipe, fittings, and adapters.

Flow Pressure — Pressure created by the rate of flow or velocity of water coming from a discharge opening.

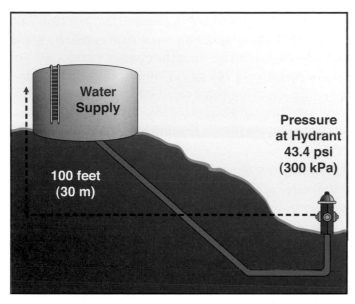

Figure 4.12 The head in this illustration is 100 ft (30m). The head pressure is 43.4 psi (300 kPa).

Figure 4.13 A pitot tube is used to measure flow pressure at the hydrant.

Friction Loss

Another concept that fire and emergency service personnel must be familiar with is **friction loss**. The fire service definition of friction loss is that part of the total pressure lost while forcing water through pipes, fittings, fire hose, and adapters. In order to effectively assess water flow for fire protection, friction loss must be understood and taken into account.

A variety of reasons that friction loss occurs include the following:

- Movement of water molecules against each other
- Condition, age, or interior surface of pipe or hose
- Couplings, valves, appliances, and fittings
- Sharp bends
- Changes in diameter of pipe or hose
- Improper gasket size

Friction loss can be measured by inserting in-line gauges in a hose or pipe. The difference in the residual pressures between gauges when water is flowing is the friction loss. For example, the difference in the pressure between the nozzle and the pumper is a good example of friction loss.

The four basic principles that govern friction loss in hose and pipes are as follows:

- ***First Principle: If all other conditions are the same, friction loss varies proportionately with the length of the hose or pipe*** — Friction loss increases as the length of hose or piping increases. Comparing two pipes that vary only in length, one 100 ft (30 m) in length and the other 200 ft (60 m) in length, both maintaining a constant flow of 200 gpm (750 L/min), will result in a friction loss of 10 psi (70 kPa) for the 100 ft (30 m) length and 20 psi (140 kPa) for the 200 ft (60 m) length **(Figure 4.14)**.

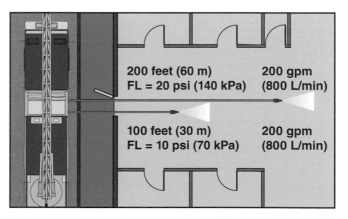

Figure 4.14 Friction loss varies proportionately with the length of pipe or hose.

- *Second Principle: When hoses or pipes are the same size, friction loss varies approximately with the square of the increase in the velocity of the flow* — Friction loss develops much faster than the change in velocity. For example, a 3-in (77 mm) pipe flowing 200 gpm (750 L/min) has a friction loss of 3.2 psi (22 kPa), while doubling the flow to 400 gpm (1 500 L/min) increases the friction loss four times to 12.8 psi (88 kPa). Tripling the original flow to 600 gpm (2 250 L/min) increases the friction loss nine times to 28.8 psi (200 kPa) (**Figure 4.15**).

- *Third Principle: For the same discharge, friction loss varies inversely to the fifth power of the diameter of the hose* — This principle illustrates the advantage of larger size hose. For example, when comparing the friction loss in a pipe that is 2½ in (65 mm) in diameter with a pipe that is 3 in (77 mm) in diameter, the friction loss in the 3 in (77 mm) pipe is 0.4 that of the 2½ in (65 mm) pipe.

$$\frac{(2½)^5}{3^5} = \frac{98}{243} = 0.4 \text{ that of the 2½ in pipe}$$

- *Fourth Principle: For a given flow velocity, friction loss is approximately the same regardless of the pressure on the water* — This principle illustrates why friction loss is the same when hoses or pipes at different pressures flow the same amount of water. For example, if 100 gpm (375 L/min) passes through a 3-in (77 mm) pipe within a certain time, the water must travel at a specified velocity (ft/sec or m/sec). For the same rate of flow to pass through a 1½-in (38 mm) pipe, the velocity must be greatly increased. Four 1½-in (38 mm) pipes are needed to flow 100 gpm (375 L/min) at the same velocity required for a single 3-in (77 mm) pipe (**Figure 4.16**).

Friction loss is affected by characteristics of hose and piping layouts such as length and diameter. Sharp bends or kinks in the hose, as well as elbows and other fittings in piping, can increase friction loss as well. Using large diameter hose, as well as removing sharp bends or kinks in fire hose, will reduce the amount of friction loss. In fire fighting operations, any extra hose should be eliminated to reduce excess friction loss.

Public Water Supply Systems

Water for human consumption comes from one of two basic sources: from a well or a public (municipal) water system. Water supply is essential for business and industry to operate and for adequate fire protection. The objectives of a public water supply system are to provide adequate and reliable services to its customers, which includes meeting fire flow and domestic needs.

Maintaining a continuous or uninterrupted supply of water for public demands is a major challenge to many municipalities because of the following conditions:

- Droughts
- Growing demands that cannot be met by the treatment plant (**Figure 4.17**)
- Lack of adequate storage capacity

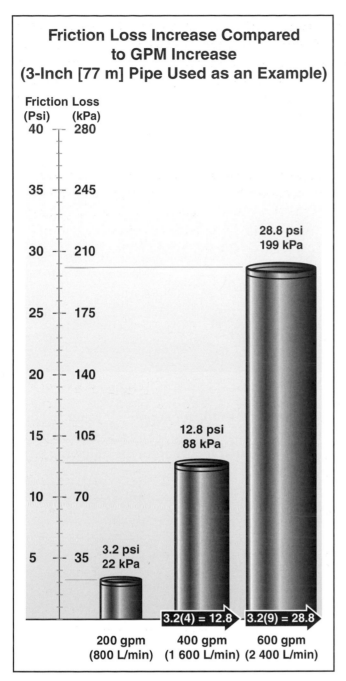

Friction Loss Increase Compared to GPM Increase
(3-Inch [77 m] Pipe Used as an Example)

Friction Loss
(Psi)   (kPa)

28.8 psi
199 kPa

12.8 psi
88 kPa

3.2 psi
22 kPa

3.2(4) = 12.8   3.2(9) = 28.8

200 gpm   400 gpm   600 gpm
(800 L/min) (1 600 L/min) (2 400 L/min)

Figure 4.15 Friction loss varies with the square of the increase in the velocity of the flow.

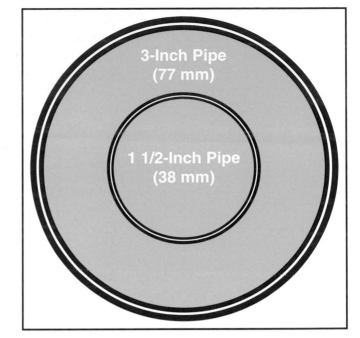

3-Inch Pipe
(77 mm)

1 1/2-Inch Pipe
(38 mm)

Figure 4.16 Doubling the diameter of the hose increases the area of the hose opening approximately four times.

Figure 4.17 Large residential and commercial growth in a municipality can quickly render its water supply capability inadequate.

- Other communities drawing water from the same supply sources such as a lake or river
- Major commercial fire or wildland/urban interface fire that exhausts the water supply **(Figure 4.18, p. 90)**
- Undetected underground leakage of the pipe distribution system
- Contamination

A municipality must recognize that the quantity of available water needs to be such that maximum daily consumption demands are satisfied at all times, even during periods of drought or after years of community growth. The water delivery system needs to expand as the municipality expands.

Figure 4.18 Major wildland fires put a tremendous strain on water supplies. *Courtesy of the Los Angeles (CA) Fire Department – ISTS.*

Public water systems provide the methods for supplying water to communities. The working parts of a water system are many and varied. While water systems are complicated, each must incorporate the following basic components:

- Sources of water supply
- Water processing or treatment facilities
- Means of moving water
- Water storage and distribution systems

Sources of Water Supply

Every water supply system has to have a supply source that is both adequate and reliable for the area being served. The primary water supply can be obtained from surface water or groundwater. Although most water systems are supplied from only one source, there are instances where both sources are used. Two examples of surface water supply are rivers and lakes. Groundwater supply can be water wells or water-producing springs.

The amount of water that a facility needs can be determined by an engineering estimate. This estimate is the total amount of water needed for domestic, industrial, and fire-fighting use. In cities, the domestic/industrial requirements far exceed those needed for fire protection. However in some locations, such as industrial facilities, the requirements for fire protection may exceed other requirements.

Water Processing or Treatment Facilities

The second component of a water supply system is the processing or treatment facility **(Figure 4.19)**. The treatment of water is a vital process. Water is treated to remove contaminants that may be detrimental to the health of those

Figure 4.19 Treatment tanks like these are commonly found in water treatment facilities.

Figure 4.20 Failure of pumps at the treatment facility can adversely impact fire fighting efforts due to decreased water supply.

who use or drink it. In addition to removing harmful elements from water, water treatment officials may add fluoride or oxygen.

The fire department's main concern regarding treatment facilities is that a maintenance error, natural disaster, loss of power supply, or fire could disable the pumping station or severely hamper the purification process **(Figure 4.20)**. Any of these situations would drastically reduce the volume and pressure of water available for fire-fighting operations. Another problem would be the inability of the treatment system to process water fast enough to meet the demand. In either case, fire officials must have a plan to deal with these potential shortfalls.

Means of Moving Water

The third component of a water supply system is a means of moving water. The following are three basic means of moving water in a municipal system:

- *Direct Pumping System* — Uses one or more pumps that take water from the primary source and discharge it into the distribution system **(Figure 4.21, p. 92)**. Failures in supply lines and pumps can usually be overcome by duplicating these units and providing a secondary power source.

- *Gravity System* — Uses a primary water source located at a higher elevation than the distribution system **(Figure 4.22, p. 92)**. The gravity flow from the higher elevation provides the water pressure. This pressure is usually sufficient only when the primary water source is located at least several hundred feet (meters) higher than the highest point in the water distribution system. The most common examples of gravity systems include a mountain reservoir that supplies water to a city below or a system of elevated tanks or water towers **(Figure 4.23, p. 93)**. These tanks provide a very reliable water supply and it is only necessary to keep the tanks full and the valves open to ensure that the water will flow when needed. However, tanks do require considerable maintenance and often require protection against freezing. They are also an attractive nuisance and are frequent targets of vandalism and mischief.

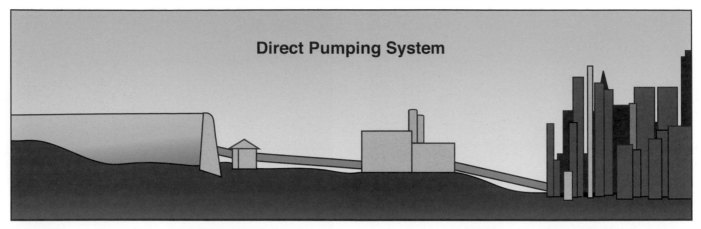

Figure 4.21 A direct pumping system is used when the water source does not have sufficient elevation to create adequate pressure.

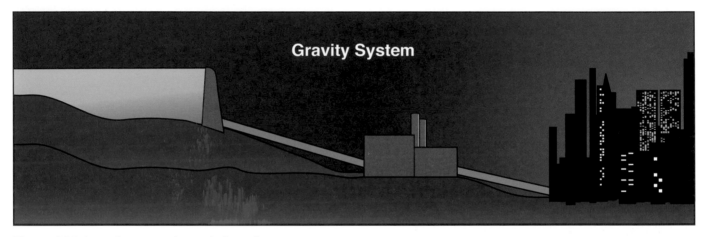

Figure 4.22 A gravity system is used where the water source is adequately elevated.

Figure 4.23 Water towers use the principles of gravity to maintain adequate water pressure. *Courtesy of the Sand Springs (OK) Fire Department.*

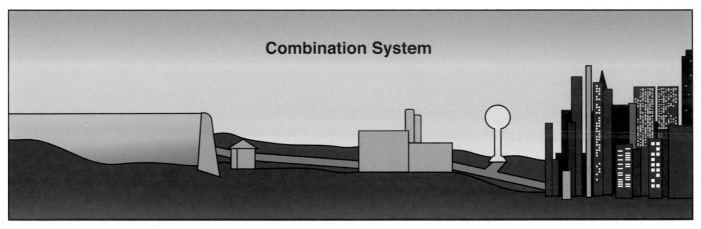

Combination System

Figure 4.24 A combination system of direct pumping and gravity is used to allow water storage during periods of low demand. When demand is increased to a level where pump capacity is exceeded, water is used from the elevated tanks.

- ***Combination System*** — Uses combinations of direct pumping and gravity systems **(Figure 4.24)**. In most cases, elevated storage tanks supply the gravity flow. These tanks serve as emergency storage and provide adequate pressure through the use of gravity. When the system pressure is high during periods of low consumption, automatic valves open and allow the elevated storage tanks to fill. When the pressure drops during periods of heavy consumption, the storage containers provide extra water by feeding it back into the distribution system. Providing a good combination system involves reliable, duplicated equipment, and proper-sized, strategically located storage containers.

Water Storage and Distribution Systems

The last components of the water supply system are the storage and distribution systems. Water storage should be sufficient to provide for domestic and industrial requirements in addition to the demands expected during firefighting operations. Such storage should also be sufficient to permit making most repairs, alterations, or additions to the system. Location of the storage and the capacity of the mains leading from this storage are also important factors. Water stored in elevated reservoirs can also ensure water supply when the system becomes otherwise inoperative.

Many industries that provide their own private water systems use storage tanks that are available to the fire department **(Figure 4.25)**. Water for fire protection may be available to some communities from storage systems, such

Figure 4.25 Some industrial sites have water storage capabilities that are made available to the fire department.

as cisterns, that are considered a part of the distribution system. The fire department pumper removes the water from these sources by drafting and provides pressure by its pump.

Water storage becomes an important fire protection feature during times of peak demand. Elevated tanks are used to stabilize or balance the pressure on a water system when demand is high **(Figure 4.26)**. As long as the pumps supplying the system can keep up with the demand, the water in the tanks will not be used. However, once the demand upon a system becomes so great that the pumps cannot keep up, the system pressure will begin to drop. When the pressure drops to a point where it cannot keep the tank full, the tank will begin to add water to the system.

The distribution system receives water from the pumping station and delivers it throughout the area served. The ability of a water system to deliver an adequate quantity of water relies upon the carrying capacity of the system's network of pipes. In order to provide adequate amounts of water for firefighting operations, friction loss in the distribution system must be taken into account. To reduce the effect of friction loss on water pressure, fire hydrants are typically supplied from two or more directions. These systems are said to have a **circulating feed** or *looped line* **(Figure 4.27)**. A distribution system that provides circulating feed from several mains constitutes a grid system **(Figure 4.28)**. A grid system should consist of the following components:

- *Primary feeders* — Large pipes (mains) with relatively widespread spacing that convey large quantities of water to various points of the system for local distribution to the smaller mains.

- *Secondary feeders* — Network of intermediate-sized pipes that reinforce the grid within the various loops of the primary feeder system and aid the concentration of the required fire flow at any point.

Circulating feed— Fire hydrant that receives water from two or more directions

Figure 4.26 Elevated tanks are used to balance the pressure on a water system during periods of peak demand.

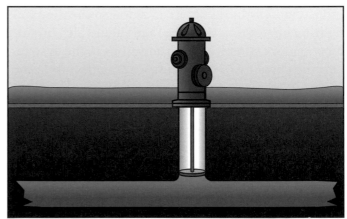

Figure 4.27 Looped fire hydrants receive water from two directions.

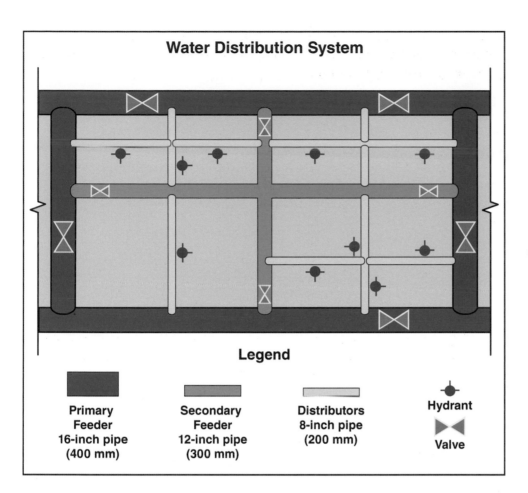

Figure 4.28 A typical grid system of water supply pipes.

- **Distributors** — Grid arrangement of smaller mains serving individual fire hydrants and blocks of customers.

- **Circulating Feed** — Fire hydrant that receives water from two or more directions.

To ensure sufficient water, two or more primary feeders should run from the source of supply to the high risk and industrial districts of the community by separate routes. Similarly, secondary feeders should be arranged in loops as far as possible to give two directions of supply to any point. This practice increases the capacity of the supply at any given location and ensures that a break in a feeder main will not completely cut off the water supply.

Private Water Supply Systems

In addition to the public water supply systems that service most communities, fire and emergency services personnel must also be familiar with the basic principles of any private water supply systems that are within their response jurisdiction. Private water supply systems are most commonly found on large commercial, industrial, or institutional properties **(Figure 4.29, p. 96)**. These systems may service one large building or a series of buildings on the complex. In general, a private water supply system exists to provide water for fire protection purposes, sanitary purposes, manufacturing processes, or a combination of these uses.

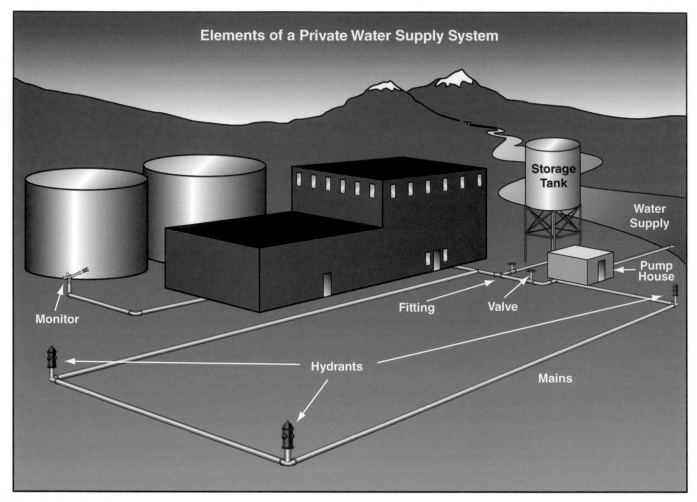

Elements of a Private Water Supply System

Storage
Tank

Water
Supply

Pump
House

Fitting

Valve

Monitor

Hydrants

Mains

Figure 4.29 Private water supply systems are commonly found on industrial properties.

The design of private water supply systems is typically similar to that of the municipal systems described earlier in this chapter. Most commonly, private water supply systems receive their water from a municipal water supply system. In some cases, the private system may have its own water supply source independent of the municipal water distribution system.

Properties may occasionally be served by two sources of water supply for fire protection: one from the municipal system and the other from a private source. In many cases, the private source of water for fire protection provides nonpotable water. When this is the case, adequate measures must be taken to prevent contamination caused by the backflow of nonpotable water into the municipal water supply system. There are a variety of backflow prevention measures that can be employed to avoid this problem. Some jurisdictions do not allow the interconnections of potable and nonpotable water supply systems. This means that the protected property is required to maintain two completely separate systems.

Many private water supply systems maintain separate piping for (1) fire protection, including hydrants, and (2) domestic and industrial uses. Separate systems are cost-prohibitive for most municipal applications but are economically feasible in many private applications. A number of advantages to having separate piping arrangements in a private water supply system include the following:

- The property owner has control over the water supply source.
- Neither system (fire protection or domestic/industrial) is affected by service interruptions to the other system.
- The facility has the ability to isolate systems for inspection, testing, and maintenance.

Fire and emergency services personnel must be familiar with the design and reliability of private water supply systems in their jurisdiction. Large, well-maintained systems may provide a reliable source of water for fire protection purposes. Small capacity, poorly maintained, or otherwise unreliable private water supply systems should not be depended upon to provide all the water necessary for adequate fire-suppression operations. Historically, many significant fire losses can be traced, at least in part, to the failure of a private water supply system that was being used by municipal fire departments working the incident. Problems such as the discontinuation of electrical service to a property whose fire protection system is supplied by electrically driven fire pumps have resulted in disastrous losses.

Water Piping Systems

Having a large capacity and reliable supply of water is of little value if the water cannot be delivered in adequate amounts and with adequate pressure to the point of use. This depends heavily upon the capacity and pressure rating of the pumps at the treatment plant and pumping stations, as well as the extent of elevated storage. The feature that has the greatest impact is the piping distribution system. The performance of the distribution system is affected by several variables, including the following:

- Piping materials
- Pipe diameter
- Piping arrangements

In addition, water main valves and fire hydrants are important parts of the water piping system, affecting the reliability of the system. Of particular importance for hydrants is their location.

Piping Materials

Piping systems are generally made of cast iron, ductile iron, asbestos cement, steel, polyvinyl chloride (PVC), plastic, or concrete. Whenever a main is installed, it must be the proper type for the soil conditions and pressures to which it will be subjected. When water mains are installed in unstable or corrosive soils, steel or reinforced concrete pipe may be used to give the strength needed. Some locations that may require extra protection include areas beneath railroad tracks and highways, areas close to heavy industrial machinery, or areas prone to earthquakes.

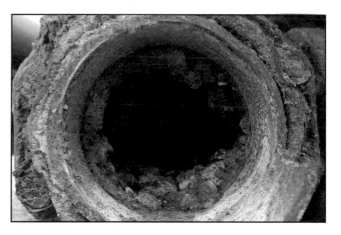

Figure 4.30 Sedimentation is an accumulation of foreign material on the interior surface of the pipe.

The internal surface of the pipe should be such that it offers the least potential for friction loss. Some materials have considerably less resistance to water flow than others. In addition to the internal surface itself, friction loss can also be increased by encrustation on the interior surface of the pipe, corrosion, or sedimentation **(Figure 4.30)**. All of these can result in a restriction of the pipe size and increase friction loss. This in turn causes a reduction in the amount of water that can be supplied by the system.

Pipe Diameters

Water distribution piping should be designed to accommodate current and future domestic and fire flow demands. It is recommended that fire hydrant supply mains be at least 6 in (150 mm) in diameter. These should be closely gridded by 8-in (200 mm) mains at intervals of not more than 600 ft (180 m). Since the water systems in many communities were constructed with little attention to fire protection demands, it is not uncommon to see 4-in (100 mm) pipe supplying fire hydrants in the older sections of many cities and towns.

Piping Arrangements

Piping arrangements are the third feature that affects water distribution system performance. Piping can be a loop system, grid system, dead-end main, or a combination thereof.

A **loop system** is an arrangement of water mains where the water will be supplied to a given point from more than one direction **(Figure 4.31)**. These are also called *circle systems*, *circulating systems*, or *belt systems*.

A **grid system**, also known as a *complex loop*, is a large piping distribution system that is characterized by multiple water pathways **(Figure 4.32)**. A well-designed grid system uses pipe sizes of 8 in (200 mm) and larger, which is one of the identifying features of a strong distribution system.

A **dead-end main** is a single pipe that extends from a looped or grid system that is supplied from one direction **(Figure 4.33, p. 100)**. This is a less desirable piping arrangement because water is supplied from only one direction and the volume of water available is typically limited.

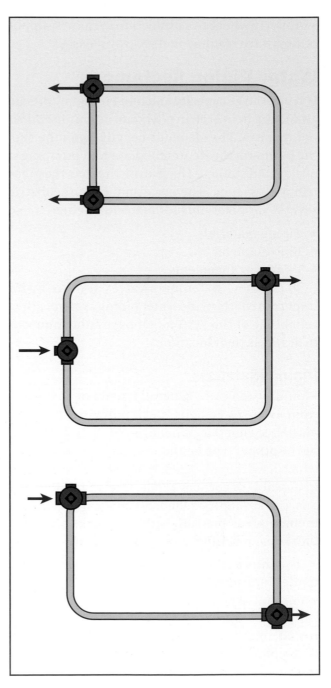

Figure 4.31 Examples of simple loops.

Water Main Valves

Valves are placed on water distribution systems to provide the ability to isolate portions of the system. This is a particularly important reliability feature. When a water main breaks, strategically closing valves necessitates only a small portion of the system being placed out of service while repairs are made **(Figure 4.34, p. 100)**. These are typically underground valves that require a special socket wrench on a reach rod, often called a *valve key*, to operate **(Figure 4.35, p. 100)**. Ideally, these valves should be located on each branch at the intersection of the mains and no more than 500 ft (150 m) apart in high-value areas. There should also be a valve on every branch feeding a hydrant so that each hydrant can be isolated during repair or replacement.

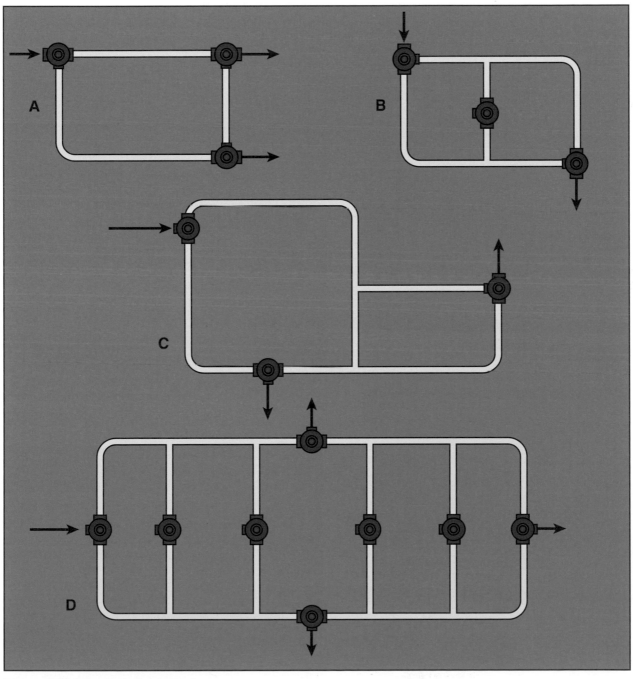

Figure 4.32 Examples of complex loops.

Since these valves are underground, access is only possible through metal valve plates located at ground level **(Figure 4.36)**. Unfortunately, these access plates are often covered with dirt or pavement, leaving valves inaccessible and their locations hidden. A good maintenance program will require accurate water maps, indicating valve locations, and yearly valve operations to make sure they are working properly and are completely open **(Figure 4.37)**.

Valves for water systems are broadly divided into indicating and nonindicating types. An **indicating valve** visually shows the position of the valve: fully open, fully closed, or partially closed. Valves in private fire protection systems are usually of the indicating type. Two common indicator valves are the **post indicator valve (PIV)** and the outside stem and yoke (OS&Y) valve **(Figures 4.38 a and b)**. The PIV is a hollow metal post that is attached to the valve housing. The valve stem inside this post has the words *OPEN* and *SHUT* printed on it so that the position of the valve is shown **(Figures 4.39 a and b, p. 102)**. The OS&Y valve has a yoke on the outside of the structure with a threaded stem that opens or closes the gate inside the valve. The threaded portion of the stem

Indicating Valve — Water main valve that visually shows the open or closed status of the valve.

Post Indicator Valve (PIV) — A type of valve used to control underground water mains that provides a visual means for indicating "open" or "shut" positions; found on the supply main of installed fire protection systems.

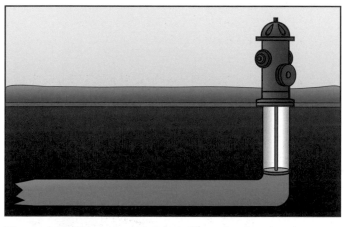

Figure 4.33 This hydrant is served by a dead-end main.

Figure 4.34 Failure of a large water main requires that the break be isolated by closing valves so that repairs can be made. *Courtesy of Tarpon Springs (FL) Fire Rescue.*

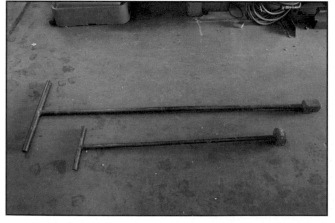

Figure 4.35 Valve keys vary in length and long keys may be needed to reach deep valves.

Figure 4.36 A typical valve access plate.

Figure 4.37 Yearly valve operations ensure that the valve is operating properly and that it is fully open.

is out of the yoke when the valve is open and inside the yoke when the valve is closed **(Figure 4.40, p. 102)**. These valves are commonly used on sprinkler systems but may be found in some water distribution system applications.

Nonindicating valves in a water distribution system are normally buried or located in utility openings. These are the most common types of valves used on most public water distribution systems. If a buried valve is properly installed, the valve can be operated aboveground through a valve box or road box.

Control valves in water distribution systems may be either gate valves or butterfly valves. Both valves can be of the indicat-

Figure 4.38a A typical post indicator valve (PIV).

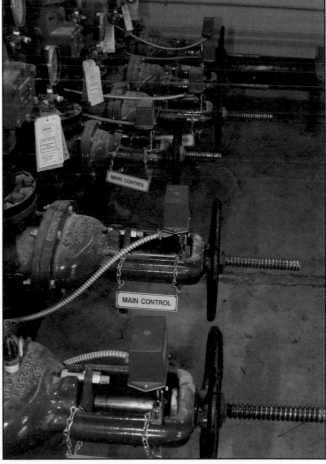

Figure 4.38b A series of outside stem and yoke (OS&Y) valves.

ing or nonindicating type. Gate valves may have a rising stem or a nonrising stem. The rising stem type is similar to the OS&Y valve. On the nonrising stem type, the gate either rises or lowers to control the water flow when the valve nut is turned by the valve key or wrench. Nonrising stem gate valves should be marked by a number indicating the number of turns necessary to completely close the valve. If a valve resists turning after fewer than the indicated number of turns, it usually means that debris or other obstructions are in the valve.

Figure 4.39a A PIV in the open position reads OPEN in the window.

Figure 4.39b A PIV in the closed position reads SHUT in the window.

Figure 4.40 In OS&Y valves, the threaded portion of the stem is out of the yoke when the valve is open.

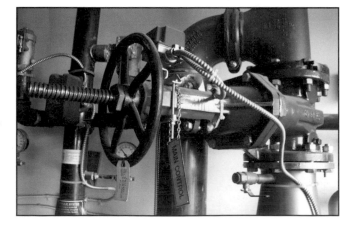

Butterfly valves are tight closing, and they usually have a rubber or a rubber-composition seat that is bonded to the valve body **(Figure 4.41)**. The valve disk rotates 90 degrees from the fully open to the tight-shut position. The nonindicating butterfly types also require a valve key. Its principles of operation provide satisfactory water control after long periods of inactivity.

A high level of friction loss will result if valves are partially closed. When valves are closed or partially closed, the condition may not be noticeable during ordinary domestic flows of water. As a result, the impairment will not be known until a fire occurs or until detailed inspections and fire flow tests are done. A fire department will experience difficulty in obtaining water in areas where there are closed or partially closed valves in the distribution system.

Fire Hydrants

Fire hydrants are the primary means with which fire and emergency services personnel obtain water for fire suppression. Hydrants are available in two basic types: dry-barrel and wet-barrel. The most common type of hydrant used in the United States is the **dry-barrel** type. As the name implies, there is no water inside the hydrant until the hydrant is turned on **(Figure 4.42, p. 104)**. This hydrant is typically used in areas where freezing temperatures are expected. The valves in the hydrant keep water out of the barrel until it is opened, and drains located at the bottom of the hydrant allow the water to drain out when the hydrant is closed. When the hydrant is fully opened, the drain holes are closed. If the hydrant is only partially opened, the drain holes will be partially open, permitting a pressurized stream of water to be discharged out of the drains beneath the ground. This drainage will erode the area at the base of the hydrant. For this reason, dry-barrel hydrants should always be completely open when in use. Dry-barrel hydrants are easily recognized by the existence of the operating nut on the *bonnet* or top of the hydrant.

A common variety of dry-barrel hydrant has two 2½ in (65 mm) outlets and one 4½ in (115 mm) outlet **(Figure 4.43, p. 104)**. The larger outlet is often called a *pumper* or *steamer* outlet because it is common for the fire department to attach its soft suction line to this outlet to feed the pumping apparatus. It is also common to see dry-barrel hydrants with only two 2½ in (65 mm) outlets and no pumper outlet. These outlets limit the capacity of the hydrant. Therefore, hydrants of this type should not be used in industrial and commercial areas of a community. There are several other combinations of outlets that may be encountered, such as hydrants having only one or two pumper outlets and no smaller outlets. However, the first two described are by far the most common.

Wet-barrel hydrants may be found where freezing temperatures are not expected, such as in southern California, Arizona, or southern Florida. Water is inside the hydrant at all times and would be subject to freezing in cold climates **(Figure 4.44, p. 105)**. The operating nuts for turning on the hydrant extend through the side of the barrel, and there must be an operating nut for each outlet on the hydrant.

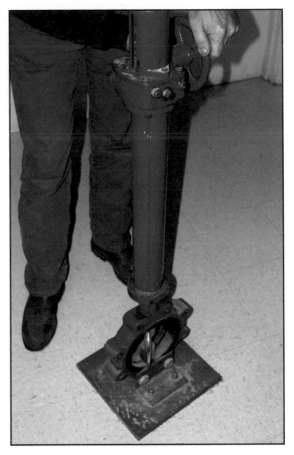

Figure 4.41 A typical butterfly valve.

Dry Barrel Hydrant   Fire hydrant that has its operating valve at the water main rather than in the barrel of the hydrant. When operating properly, there is no water in the barrel of the hydrant when it is not in use. These hydrants are used in areas where freezing could occur.

Wet-Barrel Hydrant — Fire hydrant that has water all the way up to the discharge outlets. The hydrant may have separate valves for each discharge or one valve for all the discharges. This type of hydrant is only used in areas where there is no danger of freezing weather conditions.

Whether they are wet-barrel or dry-barrel, hydrants are attached to the water main by a short piece of pipe called a *branch*. This branch should be at least 6 in (150 mm) in diameter. Branches that are 4 in (100 mm) in diameter are common on older systems, but this size limits the capacity of the hydrants. Each branch should also have a valve so that the hydrant can be isolated for repair or replacement without shutting down the water main to which it is attached.

From a maintenance standpoint, grass and bushes should be kept trimmed away from hydrants so that they remain visible and accessible. Hydrants should be flushed every year at a minimum to make sure that they work and to clean sedimentation from the pipes. It is also a good idea to measure the flow at each hydrant every year if possible. By comparing the flows at each hydrant year by year, any deterioration or obstruction of the water supply system can be determined. Testing fire flows allows for identification of varying pressures within the water system. In many communities, fire hydrants are color-coded to reflect the flow capacity of the hydrant. Refer to NFPA® 291, *Recommended Practice for Fire Flow*

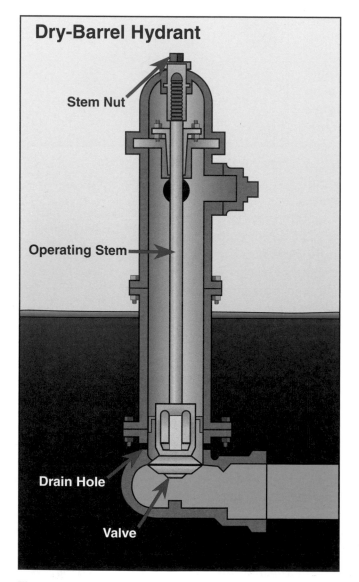

Dry-Barrel Hydrant

Stem Nut

Operating Stem

Drain Hole

Valve

Figure 4.42 Dry-barrel hydrants utilize a long operating stem to keep the water well below ground when the hydrant is not in use.

Figure 4.43 Some dry hydrants include an outlet for large diameter supply hose as well as two 2½ inch outlets.

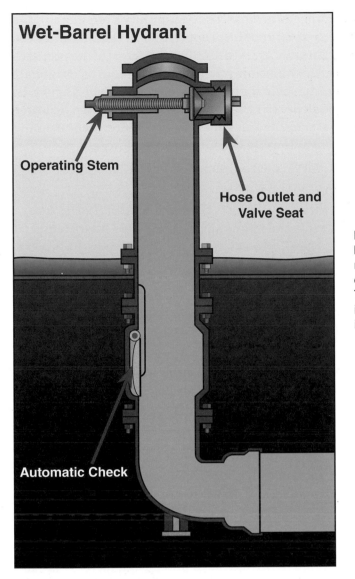

Wet-Barrel Hydrant

Operating Stem

Hose Outlet and
Valve Seat

Automatic Check

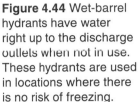

Figure 4.44 Wet-barrel hydrants have water right up to the discharge outlets when not in use. These hydrants are used in locations where there is no risk of freezing.

Testing and Marking of Hydrants, for more information. It is important for fire and emergency services organizations to have a strong working relationship with the water provider to coordinate testing and maintenance.

Hydrant Locations

In order to be effective, fire hydrants are installed at locations and with spacing consideration for convenient use by fire departments. They must also be located to meet the needed fire flows for the buildings being protected. Proper hydrant distribution and location is an important feature of an accessible water supply system. The pumper outlet should be positioned in the most advantageous location for connection to fire apparatus. In addition, the hydrant should be set high enough that the hydrant wrench can be turned a full revolution without coming in contact with the ground when removing the cap from the lowest outlet (usually the pumper outlet) **(Figure 4.45, p. 106)**.

The Insurance Services Office, Inc. (ISO), in its *Fire Suppression Rating Schedule* (FSRS), provides a method for locating fire hydrants within a community. This is based upon their Grading Schedule criteria for protection classifica-

Figure 4.45 Hydrants that are not properly maintained can be difficult to use during emergency operations. The positioning of this hydrant would make removal of the hydrant caps with a hydrant wrench difficult.

tions. The ISO evaluation procedure looks at a community's population, the number of hydrants in the system, and the property types. Communities are given credit for those hydrants that are within 1,000 ft (300 m) of the property to be protected. In addition, hydrants must deliver a minimum of 250 gpm (1 000 L/min) at 20 psi (140 kPa) residual pressure for a duration of 2 hours. In more congested areas and high fire risk areas, it is recommended that hydrants be located 300 ft (90 m) apart, one hydrant at every street intersection, in the middle of long blocks, and near the end of a long dead-end street.

The International Fire Code®, 2009 edition, Section 507, provides the following requirements for hydrant locations:

- Where a portion of a facility or building is more than 400 ft (120 m) from a hydrant, on-site fire hydrants and mains shall be provided where required by the fire code official.

- For certain occupancies (R-3 and Group U), the distance requirement is 600 ft (180 m).

- If the building is equipped with an automatic fire sprinkler, the distance requirement is 600 ft (180 m).

While there are different sources that provide information on the recommended spacing of hydrants, it is ultimately the determination of the authority having jurisdiction (AHJ) to make these decisions.

Summary

Water continues to be the most readily available and plentiful fire-extinguishing agent. In addition to knowing that water is available for fire-fighting operations, it is also important to understand how water extinguishes fire and the advantages and disadvantages of its use. In any kind of distribution system, water must be available at sufficient quantities and adequate pressures to be effective. Personnel who must inspect or use these systems need to know how water is distributed as well as any physical and design factors that can positively or negatively affect its availability. The work that is done by water purveyors and fire and emergency services personnel to keep these systems operating effectively is critical because their purpose is to save lives and property.

Review Questions

1. What four factors affect the extinguishing property of water?
2. What are the advantages and disadvantages of using water as an extinguishing agent?
3. What are the basic principles of pressure?
4. What are the different types of pressure?
5. What are some reasons that friction loss occurs?
6. What are the basic principles that govern friction loss?
7. What are the basic components of a public water supply system?
8. What are the basic means of moving water in a water supply system?
9. What variables affect the performance of a water distribution system?
10. What are the two types of fire hydrants?

Fire Pumps

Chapter Contents

chapter 5

Key Terms

FESHE Outcomes

Fire and Emergency Services Higher Education (FESHE) Outcomes: Fire Protection Systems

4. Identify the different types and components of sprinkler, standpipe, and foam systems.

Fire Pumps

Learning Objectives

After reading this chapter, students will be able to:

1. Describe types of fire pumps.
2. Describe types of pump drivers.
3. List the advantages and disadvantages of different types of pump drivers.
4. Describe the operation of a pump controller.
5. Describe pump components and accessories required for the installation of a fire pump.
6. Describe pump location and component arrangement for the installation of a fire pump.
7. Summarize primary performance criteria for fire pump testing.
8. Summarize inspection and maintenance procedures for fire pumps.

Chapter 5
Fire Pumps

Case History

In May, 1988, Los Angeles, CA firefighters responded to a reported fire in a large, 62 story high-rise office building. Construction crews were working in the building that evening to install a new sprinkler system. To do so, the contractors had to disable the 2 fire pumps that served the building's standpipe system. Minutes later, a fire was discovered. The first fire suppression crews on scene had to deal with very little water pressure coming from the building's standpipes. It was not until the fire pumps were returned to service that an effective attack on the fire could be made. The fire was eventually brought under control after 3½ hours and with the help of 383 firefighters.

Fire pumps, when used as a part of a fire protection system, play a critical role in the integrity of the system. Fire pumps are used to provide or enhance the water supply from public water systems, gravity tanks, reservoirs, lakes, ponds, or other such supply sources. Without fire pumps, many fire-suppression systems would be unable to sufficiently perform their job of fire extinguishment or containment.

A **fire pump** is a fixed pump that supplies water to a fire-suppression system. The main function of the pump is to increase the pressure of the water that flows through it. Usually a fire pump is needed to supply a sprinkler or standpipe system because the available water supply source cannot supply the pressure or volume needed.

Fire pumps can be either centrifugal, vertical-shaft turbine, or positive-displacement types. **Centrifugal pumps** are those that use an impeller and volute to create the pressure. A **vertical-shaft turbine pump** is basically a centrifugal pump with a vertical shaft that has a rotating impeller. A **positive-displacement pump** works on the principle that when pressure is applied to a confined liquid, the same pressure is equally transmitted in all directions. They capture a specific volume of fluid per pump revolution. Positive-displacement pumps may be rotary lobe, rotary vane, or piston plunger.

Fire Pump — Water pump used in public and private fire protection to provide water supply to installed fire protection systems.

Centrifugal Pump — Pump with one or more impellers that rotate and utilize centrifugal force to move the water. Most modern fire pumps are of this type.

Vertical-Shaft Turbine Pump — Fire pump originally designed to pump water from wells. Presently, it still has application when the water supply is from a nonpressurized source. Vertical-shaft pumps ordinarily have more than one impeller and are therefore multistage pumps.

Positive-Displacement Pump — Self-priming pump that utilizes a piston or interlocking rotors to move a given amount of fluid through the pump chamber with each stroke of the piston or rotation of the rotors.

Figure 5.1 The centrifugal type fire pump is most commonly used in fire protection applications.

The most common type of fire pump is the centrifugal type (**Figure 5.1**). These pumps may be installed in either a horizontal case or vertical case. The earliest fire pumps were positive-displacement types, using rotary gears or pistons.

Every pump must have a driver to operate the pump and a controller to control the function of the pump. Additionally, every fire pump must be installed in a system of piping and valves.

This chapter addresses the types of fire pumps commonly installed in businesses and industrial facilities to support existing water supplies. It discusses the pump components and accessories, common fire pump types, pump drivers, and controllers. It will also present information on testing, inspection, and maintenance of fire pumps. Criteria for the installation of fire pumps are listed in NFPA® 20, *Standard for the Installation of Stationary Pumps for Fire Protection*. Criteria for the inspection, testing, and maintenance of fire pumps can be found in NFPA® 25, *Standard for the Inspection, Testing, and Maintenance of Water-Based Fire Protection Systems*.

Common Types of Fire Pumps

Three types of fire pumps are commonly installed in all occupancy types to support existing water supplies. These are the horizontal split-case pumps, vertically mounted split-case pumps, and vertical-shaft turbine pumps. Split-case pumps are those in which the casing for the shaft and impeller is split in the middle and can be separated. This exposes the shaft, bearing, and impeller and makes for easy access for repair or replacement of parts (**Figure 5.2**). Split-case pumps can be mounted either horizontally or vertically. Vertical-shaft turbine pumps are used where the water supply is from a nonpressurized source such as a well, pond, river, or underground storage

Figure 5.2 The major parts of a centrifugal fire pump.

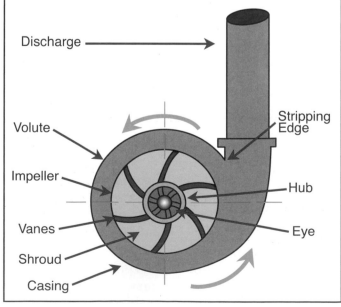

Discharge

Volute

Impeller

Vanes

Shroud

Casing

Stripping Edge

Hub

Eye

tank. These three types are discussed in the following sections. Two additional types of pumps to be mentioned include the end suction pump and the vertical inline pump.

Horizontal Split-Case Pumps

The **horizontal split-case centrifugal pump** is the most common of the split-case types and is acceptable for use where water can be supplied to the pump under some pressure. The pump is not self-priming, so it must be supplied by a public water supply system or by a tank located above the level of the pump-intake port. A pump that needs **priming** is unacceptable where automatic pump operation is required.

As its name implies, the horizontal split-case pump has its shaft oriented horizontally. A grease fitting is provided for the bearing at each end of the shaft to allow proper lubrication, which must be performed at regular intervals. There are also fiber packings on each end of the shaft, and these are inserted in the packing gland **(Figure 5.3)**. The function of fiber packings is to seal the shaft and prevent excess water leakage around the shaft. They are intended to be water cooled and lubricated so a small leakage is required. Too much or too little leakage requires tightening or loosening of the packing gland **(Figure 5.4)**.

In the horizontal split-case pump, the water pressure is increased due to the operation of a rotor inside the pump casing. This rotor is called an **impeller**. Water is fed into the center, or the eye, of the impeller from one or both sides and then thrown to the outer edges by the rotation of the impeller **(Figure 5.5, p. 114)**. As the water is forced to the outer edge of the impeller, its pressure is increased, depending on how fast the impeller is turning and the diameter and design of the impeller. The impeller turns on a shaft that is usually driven by an electric motor or a diesel engine. Though relatively rare, pumps driven by a steam turbine may be encountered. The sequence of water flow through a centrifugal pump can be seen in **Figure 5.6, p. 114**.

Horizontal Split-Case Pump — Centrifugal pump with the impeller shaft installed horizontally and often referred to as a *split-case pump*. This is because the case in which the shaft and impeller rotates is split in the middle and can be separated, exposing the shaft, bearings, and impeller.

Prime — To create a vacuum in a pump by removing air from the pump housing and intake hose in preparation to permit the drafting of water.

Impeller — Vaned, circulating member of the centrifugal pump that transmits motion to the water.

Figure 5.3 Packing glands are designed to seal the shaft and prevent excess water leakage.

Figure 5.4 Packing glands may need to be loosened or tightened to obtain the correct leakage.

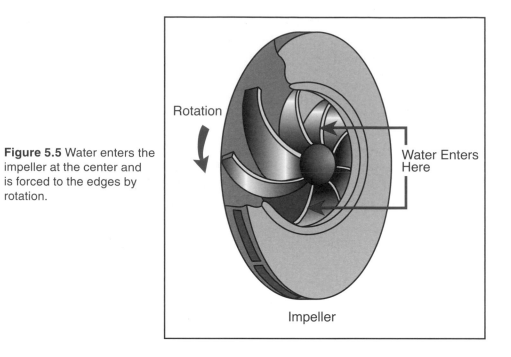

Figure 5.5 Water enters the impeller at the center and is forced to the edges by rotation.

Rotation

Water Enters Here

Impeller

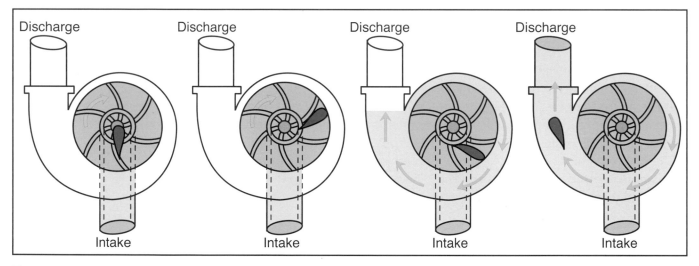

Discharge

Discharge

Discharge

Discharge

Intake

Intake

Intake

Intake

Figure 5.6 These illustrations show the path that water takes in a centrifugal pump.

Single-Stage Pump — Centrifugal pump with only one impeller.

Multiple-Stage Pump — Any centrifugal fire pump having more than one impeller.

Fire pumps are referred to as single-stage or multiple-stage. A **single-stage pump** is one with a single impeller. Most horizontal split-case pumps are single-stage pumps. **Multiple-stage pumps** have the ability to deliver higher pressures and are commonly used to supply water to standpipes in high-rise buildings.

Fire pumps that are as small as 25 gpm (100 L/min) are listed by the Underwriters Laboratories (UL) standard. While 25 gpm (100 L/min) fire pumps exist, it is rare to encounter pumps rated at less than 500 gpm (1 000 L/min). From a practical standpoint, the only time a pump smaller than 500 gpm (1 000 L/min) would be adequate for fire protection purposes would be with a small, light-hazard sprinkler system and with some standpipe and hose systems.

There is no standard pressure rating for single-stage horizontal split-case pumps, and they may be purchased with pressure ratings varying from 40 psi (280 kPa) to 290 psi (2 030 kPa). No standard fire pump is permitted to have a pressure rating less than 40 psi (280 kPa).

Vertically Mounted Split-Case Pumps

Even though the name horizontal split-case pump implies that the pump will be installed with the drive shaft in the horizontal position, there is a modern variation. In some installations where the pump is driven by an electric motor, the motor may be installed on top of the pump. This is a standard installation technique that saves floor space. In these instances, the type of pump is referred to as a **vertically mounted split-case pump**.

The vertically mounted pump is basically the same as the horizontal split-case pump and has identical applications. The impeller rotates on a shaft within the pump casing. The water is discharged in a direction perpendicular to the shaft, and the shaft rotates on ball-type bearings at each end.

Vertically Mounted Split-Case Pump — Centrifugal pump similar to the horizontal split-case, except that the shaft is oriented vertically and the driver is mounted on top of the pump.

Vertical-Shaft Turbine Pumps

The vertical-shaft turbine pump was originally designed to pump water from wells. This type of pump is still used where the water supply is from a nonpressurized source, including wells, ponds, rivers, and underground storage tanks **(Figure 5.7)**. Vertical-shaft pumps never require priming because the rotating turbines are positioned down inside the water source. These pumps ordinarily have more than one impeller and are therefore known as multistage pumps. The number of impellers is determined by the desired pressure rating — the

Figure 5.7 The impeller assembly of a vertical-shaft turbine pump is positioned inside the water source. *Courtesy of Floyd Luinstra.*

more impellers there are, the greater the pressure that will be developed. As the water exits one impeller, it enters the next, and so on until it is discharged into the fire-suppression system piping **(Figure 5.8)**.

The shaft can either be water-lubricated or oil-lubricated. A strainer is located at the bottom of the shaft to keep fish, snails, leaves, and other objects out of the pump. Each turbine sits in a slight enlargement of the casing that is referred to as the *bowl*. The bowl has a wear ring to prevent deterioration of the bowl itself.

The shaft is turned by an electric motor, a steam turbine, or a diesel engine. There is also fiber packing at the top of the shaft to seal the shaft against water leakage. The fiber packing is contained inside a packing gland, which can be either tightened or loosened to obtain the proper packing lubrication.

The vertical-shaft turbine pump has been used recently to support pressurized water supplies. In this application it is commonly referred to as a can-type installation. Essentially, a cylinder or canister is constructed with the municipal water supply feeding this canister. The vertical-shaft pump is simply set into this canister and is able to boost the pressure of the incoming water supply.

Even though the impellers of a vertical-shaft pump are located belowground inside the water source, the driver and control panel will be accessible above ground **(Figure 5.9)**. Vertical-shaft pumps are usually driven by an electric hollow-shaft motor mounted above the pump or by a diesel engine through a right-angle gear drive.

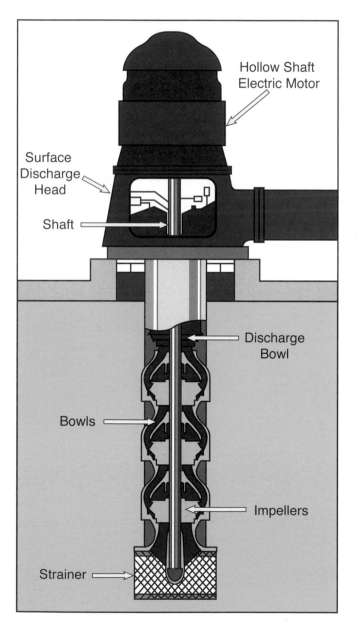

Figure 5.8 In vertical-shaft turbine applications, water moves from one impeller to the next until it is discharged from the pump.

Figure 5.9 The driver and control panel of a vertical-shaft pump are accessible aboveground.

End Suction Pumps

The end suction pump is a variation of the horizontal split-case pump design. End suction pumps are single-stage pumps that have centerline suctions and discharges. These pumps have pressure ratings from 40 to 150 psi (280 kPa to 1 050 kPa), along with flow ranges of 50 to 750 gpm (200 L/min to 3 000 L/min).

Advantages of the end suction pump are the ease of installation, simplified piping arrangement, and reduced pipe strain. The pumps are self-venting, which eliminates the need for an automatic air-release valve that is normally installed to control overheating of the pump.

Vertical Inline Pumps

The vertical inline pump is a single-stage pump designed to fit into the intake/discharge line with the driver located above the inline impeller. The advantages of a vertical inline pump are the ease of installation as a replacement pump, the small space required for the pump, and the ease of maintenance of the pump and driver. Vertical inline pumps have a capacity up to 1,500 gpm (6000 L/min) and operating pressures up to 165 psi (1 155 kPa).

Pump Drivers

The **driver** is a vital component of a fire-pump installation. Drivers are the engines or motors used to turn the pump. There are three types of drivers that are presently acceptable for use with fire pumps: electric motors, diesel engines, and steam turbines. Some earlier types of engines, such as gasoline, natural gas, and propane, can still be found in older installations. These types of engines can occasionally be found in older systems, but modern pump installations should have diesel engines when an internal combustion driver is used **(Figure 5.10)**.

The pump driver must have enough power to turn the pump at rated speed under all required load conditions, which include pumping at **churn** and pumping at 150 percent of rated capacity. A pump is said to be operating at churn or *shutoff* when it is running but all discharges are closed. The horsepower (hp) ratings are commonly found on an information plate located on the pump driver **(Figure 5.11)**. The hp varies according to the speed at which the driver

> **Driver** — Engine or motor used to turn a pump.

> **Churn** — Rotation of a centrifugal pump impeller when no discharge ports are open so that no water flows through the pump.

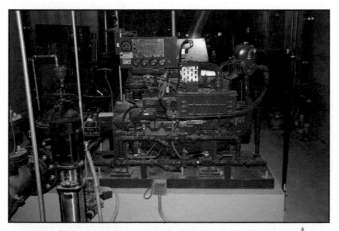

Figure 5.10 Diesel engines are the preferred pump driver if an internal combustion driver is to be used.

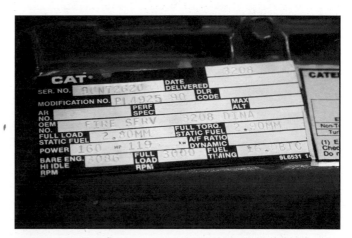

Figure 5.11 Horsepower ratings are typically found on an information plate attached to the pump driver.

is operating. For this reason, it is very important to keep a pump driver well maintained and properly adjusted. Otherwise, the pump cannot be expected to meet its performance specifications.

Electric Motors

Electric motors have long been a dependable source of power for driving centrifugal fire pumps **(Figure 5.12)**. Electric motors used on fire pumps were not originally designed for that purpose; however, all electric motors must meet the requirements of the National Electrical Manufacturers Association (NEMA). The motor must have adequate hp to drive the fire pump. The required pump hp is determined by the pump capacity (gpm), the net pressure (discharge pressure minus the incoming pressure), and the pump efficiency. For a 1,000 gpm (4 000 L/min) pump rated at 100 psi (700 kPa), a motor of about 80 hp would be needed.

Electric motors powerful enough to power fire pumps use a great deal of electricity and may require a larger electrical service to the building than would be needed otherwise **(Figure 5.13)**. For some types of pumps, such as the horizontal split-case pump, the electric motor is located on a framework beside the pump to ensure proper shaft alignment. As mentioned earlier, some split-case pumps are installed vertically and have the motor located on the top to save floor space.

One important consideration is the size of the motor relative to the pump. The motor must have adequate power to turn the pump at its rated speed so that the pump can deliver its rated flow and pressure. For this reason, a motor of ample size must be provided. The adequacy of the electric motor will become apparent during acceptance tests when the motor is required to turn the pump under various load conditions.

A substantial advantage of the electric motor is the relatively small amount of maintenance required. While electric motors are easy to maintain, they are often neglected for this very reason. Proper lubrication of the

Figure 5.12 Electric motors are a popular choice for pump drivers.

Figure 5.13 A larger electrical service than normal may be necessary for a protected structure due to the high electricity needs of an electric pump driver.

motor bearings in accordance with manufacturer's instructions is critical for reliable performance of the motor. In addition, electric pumps should be turned on weekly and permitted to run for at least 10 minutes to ensure they are operating properly.

A disadvantage of the electric motor involves reliability. Storms, fires, or other accidents involving power lines, transformers, or substations can leave motors without power and fire pumps useless. Wiring installations providing power for the electric motors and the controllers are required to comply with the provisions of NFPA® 70, *National Electrical Code*®.

Each electric motor should have an information plate that provides the current and voltage ratings, hp, revolutions per minute (rpm), and service factor. These data are useful in determining the acceptability of pump performance during the testing procedures.

Diesel Engines

The diesel engine is a common and reliable means of powering fire pumps **(Figure 5.14)**. Although diesel engines are usually more expensive than electric drivers, they may be a better choice because they do not rely on external power. While electrically driven pumps are more simple and require less maintenance, the diesel engine has proven to be the most dependable of all the internal combustion engines and is currently the only kind of internal combustion engine considered acceptable for fire protection applications.

Diesel engines are listed by UL and approved by FM Global if they are to be used for fire protection applications. This means that not all diesel engines are acceptable for driving fire pumps. If a diesel engine driver is used on a fire-pump application, the testing agencies require that the engine be equipped with overspeed shutdown devices, tachometers, oil pressure gauges,

Figure 5.14 Diesel engines are currently the only kind of internal combustion engine considered acceptable for fire protection applications.

and temperature gauges. The following aspects of a diesel engine must be considered when looking at its applicability to fire-pump use: engine power, engine requirements, cooling, and fuel storage.

Engine Power

If the engine operates at speeds that are too high, the pump can develop excessive water pressure that can damage both the engine and the pump. If the engine operates too slowly, the pump will not develop rated pressures. The engine is therefore required to have an adjustable **governor** to maintain engine speed within a 10-percent range. The governor is set to maintain rated-pump speed at maximum pump load. If the engine speed exceeds 20 percent of the rated-engine speed, the governor will shut down the pump. The shutdown device is required to send a trouble signal to the control panel until it is manually reset.

Governor — Built-in pressure-regulating device to control pump-discharge pressure by limiting engine rpm.

Engine Requirements

Several instruments must be placed on a panel that is securely fastened to the engine at an accessible location **(Figure 5.15)**. These include the following:

- *Tachometer* — Indicates the revolutions per minute (rpm) of the engine
- *Oil pressure gauge* — Indicates the pressure of the lubricating oil
- *Water temperature gauge* — Indicates the temperature of the water in the engine

Engines may be started electrically by storage batteries, compressed air, or hydraulics. Batteries are the most common means of starting diesel engines for fire pumps **(Figure 5.16)**. If batteries are used, there must be two means provided for recharging them. One may be a generator or alternator that comes with the engine, and another possibility is an automatic charger. The charger must be incorporated into the design of the controller and must be capable of fully recharging the batteries within 24 hours. The batteries should be located so that they are not subject to flooding, mechanical damage, extreme temperature variations, or vibration. They should also be readily accessible for easy servicing.

If a diesel engine is located in an environment that is subjected to flammable vapors, starting the engine through the ordinary electric starting means could create a fire or explosion hazard. For this reason, pneumatic and hydraulic starting is available. This is accomplished by forcing a compressed gas or water through a turbine that turns the engine.

For safety reasons, exhaust fumes from diesel engines should be piped to the outside of the building. Exhaust piping should not be located close to combustible materials. The exhaust system should be inspected regularly for leaks.

Cooling System

Diesel engines are water-cooled and make use of a closed-circuit-type cooling system. The basic components of the system include a water pump driven by the engine, a heat exchanger, and a reliable device for regulating the water temperature in the engine jacket. The heat exchanger works by taking water from the discharge side of the pump to cool the water in the engine. The exchange of heat from the engine water to the pump discharge water takes place in the heat exchanger. The provision of water from the discharge side of the pump requires a special piping arrangement, including a bypass line, valves, a pressure regulator, and strainers.

Fuel Storage

Diesel fuel is not as volatile as gasoline, but it can still be dangerous and must be handled carefully. Safe storage, transmission, and adequate quantities must be provided **(Figure 5.17)**. Any exposed fuel lines must be protected against mechanical damage. Fuel piping should be rigid except where the fuel line connects to the engine. At that point, flame-resistant flexible hose is required.

For environmental protection, containment should be provided to prevent runoff from any leakage from the tank. This is commonly accomplished through construction of a small containment dike around the base of the tank. Codes and ordinances usually require the containment volume to be at least as large as the total volume of the tank. Other containment methods include sloping

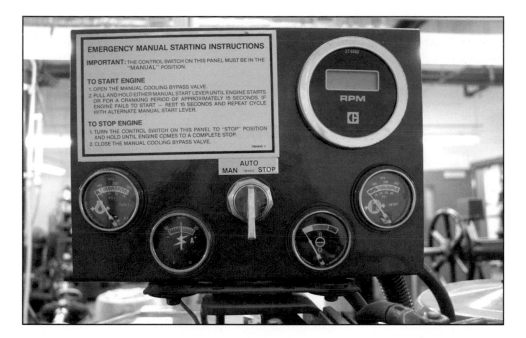

Figure 5.15 The instrument panel of a diesel pump driver typically includes the tachometer, an oil pressure gauge, and a water temperature gauge.

Figure 5.17 Diesel for pump drivers must be stored safely and in adequate quantities.

Figure 5.16 Batteries are the most common means of starting diesel pump drivers, but must be provided with a means for recharging them.

the floor to channel leaks to a containment basin or placing a sill around the pump room. There are numerous other considerations with regard to system design that are outside the scope of this manual. For more information, please reference IFSTA's **Fire Inspection and Code Enforcement** manual, 7th Edition.

Steam Turbines

Although not common, some fire pumps are driven by steam. NFPA® 20 lists steam turbines as an acceptable type of pump driver. Some of the old installations used piston-type reciprocating steam engines. The only suitable application of the steam-driven pump is when an uninterrupted supply of steam is available in sufficient quantities and at sufficient pressure. Otherwise, economic considerations would dictate use of the electric- or diesel-driven equipment.

The steam turbine driver provides steam pressure to drive both horizontal and vertical split-case pumps directly. The steam turbine driver should be connected directly to the fire pump. In these applications, steam turbines should be equipped with a safety valve to relieve high steam pressure in the casing. Steam and exhaust chambers should be equipped with condensation drains to allow for drainage of the condensation from the equipment.

Pump Controllers

A **pump controller** is another critical aspect of a fire-pump installation. Pump controllers are the control panels used to switch the pumps on and off and control their operation (**Figure 5.18**). These panels govern the operation of the pump. The controller can be designed to operate the pump automatically by use of microprocessors, simple electronic circuits, or manual operation (**Figure 5.19**). Electric pump controllers may or may not have a power cut-off between the power company connection and the controller. Therefore, only qualified personnel should service the pump controller because of the high-voltage wiring it contains. The following sections address the controllers for electric motor-driven pumps, diesel pumps, and steam turbines.

Figure 5.18 Pump controllers are the control panels that turn fire pumps on and off and control their operation.

Figure 5.19 The operational components of pump controllers can range from simple to extremely complex.

Controllers for Electric Motor-Driven Pumps

The controller for an electrically driven pump should be tested and listed for fire protection use by one of the nationally recognized testing laboratories. The controller should be located inside the pump room and as close to the pump as possible. It should be protected from water discharge, and all current-carrying parts of the controller should be at least 12 inches (300 mm) off the floor. The main parts of the controller for electric drivers are as follows, and each part can be found by associated number in **Figure 5.20**:

1. *Circuit breaker* — The function of the circuit breaker is to provide overcurrent protection by opening if too many amperes (amps) are being drawn. The circuit breaker should be accessible and operable from the outside of the controller. Proper load rating of the circuit breaker is important. It should allow at least 114 percent of the rated full-load current without tripping and also permit normal starting of the motor without tripping.

2. **Isolation switch** — This is located between the power supply and the circuit breaker. In some controllers, the isolation switch and the circuit breaker are interlocked so that the isolation switch cannot be operated with the circuit breaker closed. In other cases, the operating handle of the isolation switch is equipped with a spring latch that requires the use of both hands to operate. The isolation switch lever is always located outside the control panel.

3. **Pilot lamp** — Every controller has a pilot lamp. It may be found in different locations. The purpose of the pilot lamp is to indicate when power is available to the pump control panel.

4. **Manual start and stop button** — These are standard controller components and are located on the outside of the controller enclosures. Manual starting and stopping of the pump can be accomplished by using these buttons.

5. **Emergency start lever** — This switch can be latched in the operating position and provides for continuous nonautomatic operation. This operation is independent of the timer or automatic starting mechanisms.

6. **Running period timer** — The running period timer is used to shut off the motor after the situation has returned to normal. The timer should be set for at least 10 minutes. Therefore, if a pump is automatically controlled, it will operate for at least 10 minutes after being turned on unless manually shut down.

7. **Pressure switch** — This is the most commonly used method for automatically turning on a fire pump. It is set to close the circuitry and turn on the fire pump when water pressure on the system drops below the switch setting. Such a pressure drop is usually caused by a sprinkler opening or by hoselines being operated somewhere in the system. The pressure switch is adjustable with high and low settings. When the pressure on the system drops to the low-pressure setting, the pump turns on. When the pressure is restored to normal, the high pressure setting turns the pump off.

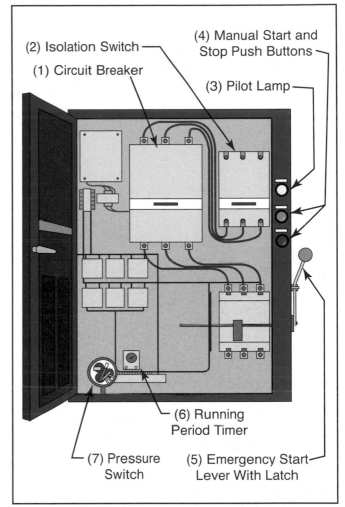

Figure 5.20 Electric fire pump controllers all have the same basic features.

Consistent pressure fluctuations may require the use of a pressure-maintenance pump, sometimes referred to as a *jockey pump*, to prevent false pump starts. It should be noted that if the pump is the sole source of water supply to the fire protection equipment, the pump should be set for automatic start and manual shutdown.

Sometimes a pump is used to supply a deluge system or dry pipe system where it is necessary to turn on the pump independent of a pressure drop. In this case, relays can be used to turn the pump on when a signal is received from a fire alarm control panel (FACP).

If the pump installation is the only source of water supply pressure for a sprinkler system or standpipe system, the automatic start controller must be wired for manual shutdown. This is done to ensure that the pump will be turned off *only* after a fire emergency is over. In addition, if the pump room is not constantly attended, audible or visible alarms should be transmitted to an attended location to signal that the pump is running or to signal that the power supply to the pump has been interrupted.

Some pumps may not be equipped with automatic controllers. If manual controllers are used, there will be both a manually operated electric switch to turn the pump on and off, in addition to a mechanical control consisting of a lever or handle that can be latched in the ON position.

Diesel Engine Controllers

The controllers for diesel engines are not interchangeable with the controllers for electric motors **(Figure 5.21)**. Electric motor controllers are designed to start the pumps by closing high voltage circuits. For diesel controllers, the main function is to close the circuit between the storage batteries and the engine starter motor.

Two important features to note are the alarm and signal devices located on the controller itself. If the controller is automatic, a pilot light will indicate when the controller is in the AUTOMATIC position. Separate lights and a common audible alarm are also required to indicate special occurrences such as the following:

- Low engine oil pressure
- High engine coolant temperature
- Failure of the engine to start automatically
- Engine shutdown due to overspeed
- Battery failure
- Low pressure in the storage tanks when air or hydraulic starting is used

The controller also requires separate lights for battery-charger failure, but this does not require an audible alarm.

If the pump room is not constantly attended, alarm signals should be transmitted to a constantly attended location. The alarm will indicate when the engine has started, when the controller has been turned off or turned to manual operation, and when trouble exists with the engine or controller.

The controller may also be equipped with a pressure-recording device. The recorder is usually required to run continuously for at least 7 days without resetting or rewinding. The recorder's chart drive should be spring-wound, alternating current (AC) electric powered with spring-driven backup, or air-powered.

The provision for automatic starting of the engine due to water pressure or fire protection device activation is similar to the provision for the electric motor controller. If the pump installation is the sole supply to standpipes

Figure 5.21 While they may appear similar, controllers for diesel engines are not interchangeable with controllers for electric drivers.

or sprinkler systems, the automatic start controller should be wired for manual shutdown. In addition, the automatic controller usually must be arranged to automatically start the engine every week to ensure reliability in engine starting. If the controller is designed to automatically shut down the engine when system conditions return to normal, it should provide running time of at least 30 minutes before shutting the engine off. However, if the overspeed governor operates, the controller will shut off the engine without a time delay.

Steam Turbine Controllers

Steam turbines are equipped with a speed governor that will maintain the rated speed for the maximum pump load. There will also be an independent emergency governing device. This emergency device will be configured to shut off the steam supply to the turbine at 20 percent higher than the rated pump speed. The speed governor should be able to adjust the speed of the pump so that it is operating at approximately 5 percent above to 5 percent below the rated speed of the pump when it is running at its rated pump load. A steam pressure gauge should be visible on the side of the governor.

Pump Components and Accessories

While each pump installation may be different, there are basic components that are common to all installations. These fit together to make a complete fire pump installation. Pump components and accessories include piping and fittings, relief valves, test equipment, pressure maintenance pumps, and gauges.

Piping and Fittings

All underground piping that supplies a fire pump must be installed and tested in accordance with NFPA® 24, *Standard for the Installation of Private Fire Service Mains and Their Appurtenances.* These pipes may be constructed of materials including steel, plastic, asbestos cement, and ductile iron.

The aboveground pipe must be constructed of steel **(Figure 5.22)**. The steel pipe can be joined together by screwed or flanged fittings or with grooved fittings. The pipe can also be welded together. Suction piping must also be installed and tested in accordance with NFPA® 24. A control valve of the outside stem and yoke (OS&Y) type must be located in the suction line for control of the water supply.

The proper sizing of both suction and discharge piping is important. If piping size is too small, the pump cannot perform at its rated capacity. The required size depends upon the rating of the pump.

In most cases, the minimum sizes are the same for both the suction side and the discharge side. However, for a 1,000 gpm (4 000 L/min) pump, the minimum for

Figure 5.22 All aboveground piping must be constructed of steel.

the suction side of the pump is larger than the discharge side. Although this is not a common requirement, in practice it is common to see the suction pipe larger than the discharge pipe. The larger piping reduces friction loss and increases the pressure available at the pump.

When the size of the suction pipe is different from the inlet port of the pump, a reducer will need to be installed in the suction line. The reducer is required to be of the eccentric type, not a concentric type (**Figure 5.23**). The reason that the reducer must be an eccentric type is to eliminate the possibility of air pockets. Air becoming entrapped in the water stream can reduce the efficiency of the pump, and can even cause damage due to **cavitation**. The eccentric reducer is to be installed with the flat side on the top.

According to NFPA® 20, no device should be installed that will restrict water flow on the suction side with a horizontal split-case pump, including backflow prevention devices (**Figure 5.24**). These devices must be located on the discharge side of the pump. In addition, a strainer may be found on the suction side; it is most likely to be used where nonpotable water is supplied. Strainers can also restrict the flow to the pump, especially if they are undersized or dirty.

Relief Valves

Historically, if a pump was driven by a variable-speed driver such as a diesel engine, a **pressure relief valve** was required in the installation (**Figure 5.25**). More recent versions of NFPA® 20 require these large relief valves only if the pressure at churn is high enough to damage system components. The purpose of this relief valve is to prevent pressures from reaching levels that are high enough to damage system piping or fittings. It is possible for an internal combustion engine to lose adjustment and develop excessive speeds. Pump pressure is related to the square of the pump speed. For example, doubling the revolutions per minute increases the pressure developed fourfold. Therefore, the relief valve is provided to open and discharge water to a drain if the pressure becomes excessive. The relief valve should be located between the pump and the discharge check valve. The valve should discharge into an open drain and in a manner so that the water flow can be visually detected.

The size of the relief valve and its discharge line depends upon the rating of the pump. If the pump was chosen with a proper pressure rating, the relief valve should be set to open a little above the normal discharge pressure when the pump is operating at churn.

A small relief valve, called a **circulation relief valve**, may be present on electrically driven pump installations. It is designed to open and provide enough water flow into and out of the pump to prevent the pump from overheating when it is operating at churn against a closed system. All pumps are required to have this circulation relief valve, except for the diesel engine driven pump where engine-cooling water is taken from the pump discharge.

When the pump is operating at churn, the circulation relief valve should be set to flow a full stream. It should then close off when water begins to flow in the system. The discharge piping should be set to discharge into a drain where the flow can be inspected.

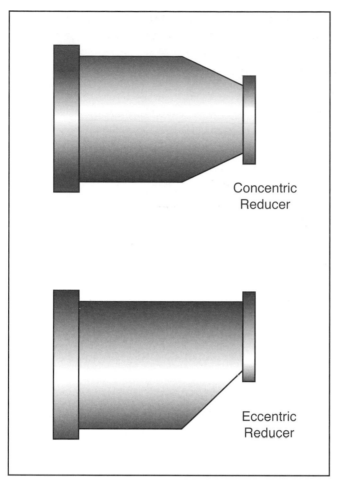

Concentric
Reducer

Eccentric
Reducer

Figure 5.23 An eccentric reducer is used in the suction line to eliminate the possibility of air pockets.

Figure 5.24 Backflow prevention devices are common and are designed to prevent water from traveling back into the municipal water supply.

Figure 5.25 Pressure relief valves are included in the installation to prevent pressures from reaching levels that would damage piping or fittings.

Test Equipment

Every pump is required to have the necessary components for testing the installation. The most prominent of these components is the test manifold **(Figure 5.26)**. The test piping should be connected to the pump discharge line between the check valve and the indicating control valve. There should also be an indicating control valve in the test piping. This piping should terminate in a hose valve header located outside the building. The hose valve header should be equipped with the proper hose connections and a shutoff valve for each connection. This manifold and test header will permit water to flow from the pump installation through hoselines and nozzles for test purposes. The flow from the nozzles is measured using pitot tubes.

The required size of the hose header supply pipe and the number of hose valves required depends upon the rating of the pump. It is often possible to estimate the rating of the pump by counting the hose connections. There is usually one 2½ in (65 mm) hose

Figure 5.26 A typical test manifold.

Table 5.1 Required Number of Test Valves				
Pump Rating		Number of Hose Valves	Test Pipe Size	
(GPM)	(L/min)		(Inches)	(mm)
500	2 000	2	4	100
750	3 000	3	6	150
1,000	4 000	4	6	150
1,250	5 000	6	8	200
1,500	6 000	6	8	200
2,000	8 000	6	8	200

Figure 5.27 Metering devices measure the gallons per minute (L/min) delivered by the pump.

connection for each 250 gpm (1 000 L/min) of pump rating. For example, a 500 gpm (2 000 L/min) pump usually has two hose connections and a 750 gpm (3 000 L/min) pump usually has three hose connections. This rule of thumb becomes less reliable with 1,250 gpm (5 000 L/min) and 2,000 gpm (8 000 L/min) pumps **(See Table 5.1)**.

More modern pump installations do not have the test headers and hose valves. Instead, these installations are equipped with a metering device that can be used to measure the gpm (l/min) delivered by the pump **(Figure 5.27)**. This is acceptable, but the meter line should discharge to the outside or back to the water supply source, not directly back to the suction side of the pump. Circulating back to the suction side of the pump will not enable the condition of the suction supply to be evaluated.

Pressure Maintenance Pumps

Sometimes there is enough leakage in the fire protection system or enough fluctuation in the pressure of the water supply to the pump to cause the automatic controller to turn the pump on periodically in nonemergency situations. This can be a nuisance if it happens frequently, particularly in installations that require manual shutdown after automatic operation. A **pressure maintenance pump** is used to prevent these false starts **(Figure 5.28)**.

Pressure Maintenance Pump — A pump used to maintain pressure on a fire protection system in order to prevent false starts at the fire pump.

A pressure maintenance pump is a small-capacity, high-pressure pump used to maintain constant pressures on the fire protection system. This pump takes suction from the fire pump suction line and discharges into the fire pump discharge line on the system side or downstream side of the indicating control valve. The pressure maintenance pump should have adequate capacity to keep up with any leaks in the system. It should be small enough, however, that any demand on the fire protection system, even a single sprinkler opening, will result in the operation of the main fire pump.

The pressure rating of the pressure maintenance pump should be high enough to maintain the desired fire protection system pressure. The pressure switch in the fire pump controller should be set to correspond to the system pressure maintained by the pressure maintenance pump. Thus, when the pressure maintenance pump cannot maintain the system pressure due to a demand on the system, the fire pump controller will activate the fire pump

Figure 5.29 Because of their importance in preventing wear on the pump driver, pressure maintenance pumps should be inspected regularly.

Figure 5.28 Pressure maintenance pumps, also called jockey pumps, are used to maintain constant pressures without having to activate the main pump.

(Figure 5.29). The pressure maintenance pump installation requires the provision of a check valve in the discharge pipe from the pressure maintenance pump as well as indicating control valves. In some cases, pressure relief valves are also required.

Pumps that are automatically controlled should be provided with an automatic air release. This air release is a float-type device that automatically closes when the pump casing fills with water and allows the release of any air that may be trapped inside the pump. Air inside the pump can reduce efficiency and can even cause damage to the pump.

Gauges

Some pumps are required to have a single gauge. Others, like vertical-shaft pumps, are required to have two gauges. On those pumps having two gauges, one should be located near the discharge port of the pump and the other near the intake. These gauges should be a certain size and diameter and capable of registering pressures of at least 200 psi (1 400 kPa) or twice the rated pressure of the pump, whichever is greater. However, gauges of all sizes and with various pressure ranges are likely to be encountered. The suction gauge must be a compound gauge that registers both positive and negative gauge pressures **(Figure 5.30, p. 130)**. Often, an ordinary gauge will register only positive pressures.

For those pumps requiring only a single gauge, it will be located on the discharge line. Special attention should be given to these pressure gauges. Inaccurate gauges can invalidate test results and give a false indication of how well a pump is performing. During the testing of a pump, the gauge needle may vibrate so widely that an accurate reading is very difficult to obtain. If this is the case, the use of a liquid-filled gauge can eliminate the vibration problem **(Figure 5.31)**.

Figure 5.30 Suction gauges must be able to register both positive and negative pressures.

Figure 5.31 Liquid filled gauges can be used in applications where there is significant vibration.

Pump Location and Component Arrangement

Considerable planning is necessary when selecting the appropriate pump, components, and accessories for a fire protection system. This planning is also necessary when selecting a physical location and arranging these items.

Location and Protection of Fire Pumps

The pump, driver, and controller of a fire pump are required to be protected against possible interruption of service caused by damage due to explosion, fire, flood, earthquake, windstorms, freezing, vandalism, or other adverse conditions. At a minimum, outdoor equipment should be shielded by a roof or deck. Fire pump units located outdoors or in buildings other than the building being protected must be located at least 50 feet (15 m) away from the protected building. Indoor fire pump units must be separated from all other areas of the building by fire-rated construction **(Figure 5.32)**.

The fire pump room must be provided with heating equipment capable of maintaining a minimum of 40°F (4°C). If a fire alarm system is present, the temperature of the pump room must be monitored. The pump room must be provided with proper ventilation. Artificial lighting must be provided in the fire pump location along with emergency lighting.

Component Arrangement

The major components of a pump installation have been identified previously in the chapter. Shown in **Figure 5.33** is a line drawing of a typical fire pump installation using a horizontal split-case pump. With this type of pump, the indicating control valves on the supply side of the fire pump and pressure maintenance pump must be OS&Y type. The other control valves shall be of the indicating type.

It may seem as though there are many indicating control valves required in a pump installation. The reason for so many valves is the ability to isolate any component, such as the pressure maintenance pump or the check valve, in the bypass line **(Figure 5.34)**. This enables a component to be repaired or even removed without having to shut off the pump installation. **Figure 5.35, p. 132** shows a horizontal split-case pump installation taking water from a storage tank with a positive head.

With a vertical-shaft pump, the components on the discharge side of the pump are essentially the same as they are for the horizontal pump **(Figure 5.36, p. 133)**. There is a relief valve (if needed), the **check valve**, and the test and discharge pipes with their indicating control valves. A vertical-shaft pump often takes suction from a wet pit. This type of installation would ordinarily exist where the water supply source is a pond, lake, or river. An important feature is the double screening required between the pond and the pit that is designed to keep foreign matter out of the pump. The screens should be removable for easy cleaning.

Check Valve — Automatic valve that permits liquid flow in only one direction. For example, the inline valve that prevents water from flowing into a foam concentrate container when the nozzle is turned off or there is a kink in the hoseline.

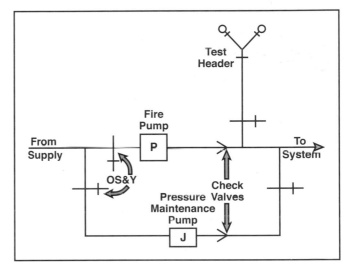

Figure 5.33 A line drawing of a horizontal split-case pump installation.

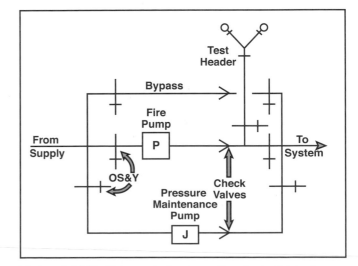

Figure 5.34 Multiple control valves allow for individual components to be isolated in the system should repairs or replacement be necessary.

Figure 5.32 This fire pump room is constructed using fire resistant materials.

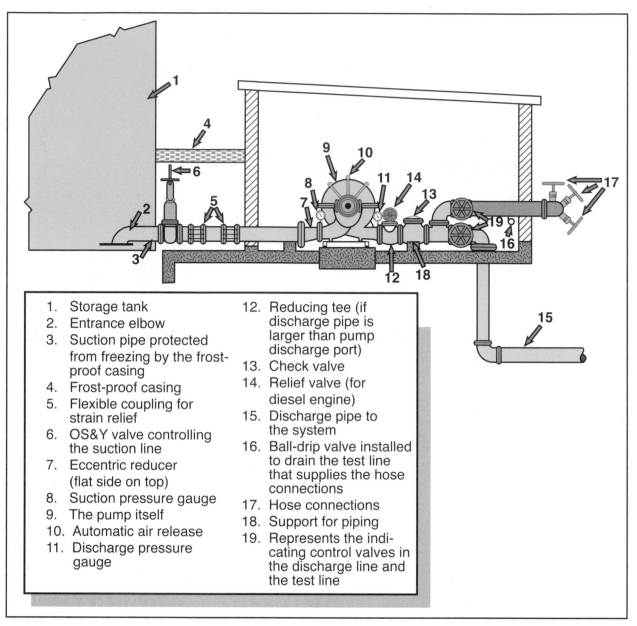

1. Storage tank
2. Entrance elbow
3. Suction pipe protected from freezing by the frost-proof casing
4. Frost-proof casing
5. Flexible coupling for strain relief
6. OS&Y valve controlling the suction line
7. Eccentric reducer (flat side on top)
8. Suction pressure gauge
9. The pump itself
10. Automatic air release
11. Discharge pressure gauge
12. Reducing tee (if discharge pipe is larger than pump discharge port)
13. Check valve
14. Relief valve (for diesel engine)
15. Discharge pipe to the system
16. Ball-drip valve installed to drain the test line that supplies the hose connections
17. Hose connections
18. Support for piping
19. Represents the indicating control valves in the discharge line and the test line

Figure 5.35 A horizontal split-case fire pump installation with water supply under a positive head.

Testing, Inspection, and Maintenance of Fire Pumps

Fire pumps are extremely important in fire-suppression systems because they must be relied upon to provide adequate water flows when needed. Due to their importance, fire pumps must be properly tested, inspected, and maintained at required intervals in order to ensure proper operation. The following sections address important concepts of fire pump testing, inspection, and maintenance.

Fire Pump Testing

The primary performance criteria for standard fire pumps are contained in NFPA® 20. The FM Global data sheets also contain information on fire pumps, comparable to NFPA® 20. In order to be considered standard under

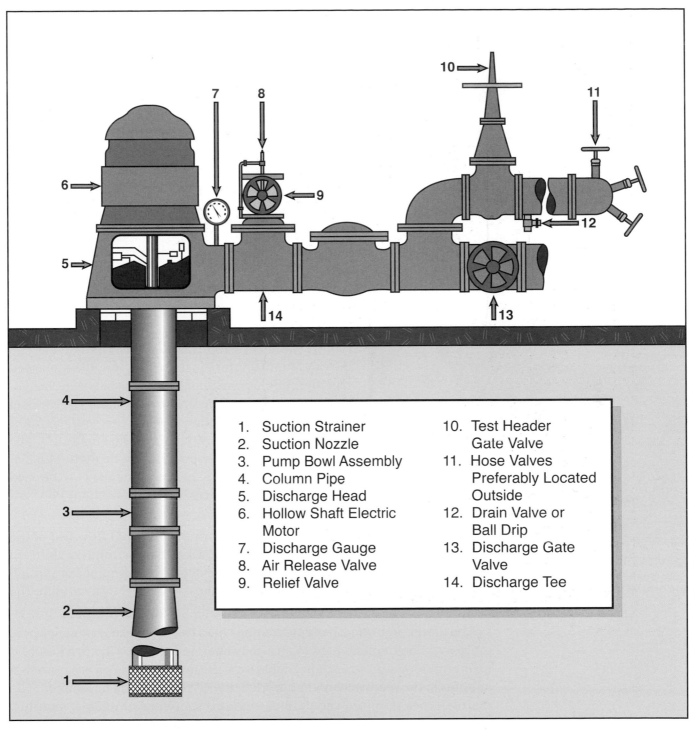

Figure 5.36 A vertical shaft, turbine type fire pump.

1. Suction Strainer
2. Suction Nozzle
3. Pump Bowl Assembly
4. Column Pipe
5. Discharge Head
6. Hollow Shaft Electric Motor
7. Discharge Gauge
8. Air Release Valve
9. Relief Valve
10. Test Header Gate Valve
11. Hose Valves Preferably Located Outside
12. Drain Valve or Ball Drip
13. Discharge Gate Valve
14. Discharge Tee

the provisions of NFPA® 20, a new fire pump must be capable of satisfying the following three test points:

- The pump must not develop more than 140 percent of its rated pressure when operating against a closed system or at churn. However, it should be noted that all horizontal pumps manufactured before 1987 were required not to exceed 120 percent of the rated pressure at churn.

- The pump must operate at minimum of 100 percent of its rated capacity at 100 percent of its rated pressure.

- The pump must operate at 65 percent of its rated pressure while delivering 150 percent of the rated volume in gpm (L/min) or the predetermined maximum volume.

Refer to Appendix A of this manual for more detailed information on the testing of fire pump installations.

Inspection and Maintenance of Fire Pumps

Fire pumps that fail to operate when needed are likely to result in catastrophic losses. The way to prevent pump failure is to regularly ensure that pump installations are in good operational condition.

Model codes require that pumps be operated at least weekly. It is not necessary to actually discharge water during the weekly startup. A diesel engine shall be operated for at least 30 minutes, while an electric motor shall be operated for 10 minutes.

Figure 5.37 Electric pump drivers are often neglected due to low maintenance requirements, but should be given the same attention as other system components.

Electric motor driven pumps are often neglected **(Figure 5.37)**. This is partly due to their simplicity and to the fact that very little maintenance is required. Diesel engines require more care and maintenance than do electric motors. Manufacturer's instructions should be followed regarding oil changes, and the interval between changes should never exceed 1 year. Batteries should be checked regularly to ensure that they are fully charged. Fuel tanks should be kept free of water and foreign materials. The temperature in pump rooms should be maintained at 70°F (21°C). If necessary, automatic heaters can be used to keep the engine warm. The engine should be kept clean, dry, and well lubricated.

NFPA® 25 requires that an annual flow test of the pump assembly be performed to determine its ability to continue to attain satisfactory performance at shutoff, rated flow, and peak loads. Annual flow tests allow the performance of the pump to be compared year by year.

The manufacturer's recommendations for a preventive maintenance program must be implemented to ensure proper operation of the fire pump. If no maintenance program is available from the manufacturer, a program that adheres to the requirements as outlined in NFPA® 25 should be adopted. The maintenance program should incorporate a sequence of weekly, monthly, quarterly, and annual tests, along with recommended maintenance that will ensure the continued satisfactory performance of the fire pump assemblies.

Although some types of bearings do not require grease, special attention must be given to clean and lubricate all bearings. The appropriate quantity and the correct type of lubricant must be applied.

Some older pumps may have packing that needs to be checked and adjusted monthly. Others may be equipped with mechanical seals rather than fiber packing. If this is the case and the mechanical seals are in good condition, no leakage should be visible. In addition to the packing and seals, it is a good practice to recheck the pump alignment regularly. Misalignment of the pump drive couplings can cause excessive vibration and pump damage.

During weekly inspections, it is also important to make sure that the pump room is kept clean, dry, and free of combustible materials. The weekly inspection is a good time to ensure that all control valves that are supposed to be open are open and are supervised in the OPEN position by padlock and chain, electronic equipment, or tag and seal. This is also the appropriate time to make sure that water tanks and diesel fuel tanks are full.

Personnel will eventually encounter a situation where the pump installation is not operating properly. NFPA® 25 provides a troubleshooting checklist in its Annex C that can help in identifying the causes of pump problems.

Summary

Fire pump installations are vital parts of a total fire protection system. Their proper selection, installation, and maintenance can be the difference between a quick recovery after a fire and a catastrophe that could destroy people, property, and jobs.

Fire pumps may be used to provide water supplies where existing supplies are deficient. Modern pumps must be able to provide dependable service. In order to do this, they must be properly installed, tested, and maintained.

Review Questions

1. What are some common types of fire pumps?

2. What are the three types of drivers acceptable for use with fire pumps?

3. What is a pump controller?

4. What are the main parts of a controller for electric drivers?

5. What are the basic pump components and accessories necessary for a fire pump installation?

6. What is the purpose of a relief valve?

7. What is the purpose of a pressure maintenance pump?

8. What are the requirements for the fire pump room?

9. What are the three test points that a new fire pump must meet?

10. How can pump failure be prevented?

Automatic Sprinkler Systems

Chapter Contents

chapter 6

Key Terms

FESHE Outcomes

Fire and Emergency Services Higher Education (FESHE) Outcomes: Fire Protection Systems

1. Explain the benefits of fire protection systems in various types of structures.

4. Identify the different types and components of sprinkler, standpipe, and foam systems.

10. Discuss the appropriate application of fire protection systems.

Automatic Sprinkler Systems

Learning Objectives

After reading this chapter, students will be able to:

1. Discuss the benefits of automatic sprinkler systems and reasons for their installation.
2. Describe components of an automatic sprinkler system.
3. Discuss temperature ratings, response times, and deflector components of sprinklers.
4. Describe types of specialty sprinklers.
5. Discuss wet-pipe and dry-pipe sprinkler systems.
6. Describe deluge systems and preaction systems.
7. Discuss residential sprinkler systems.
8. Discuss sprinkler systems in storage facilities.
9. Summarize considerations for the testing and inspection of sprinkler systems.

Chapter 6
Automatic Sprinkler Systems

Case History

In 2000, a fire broke out in the kitchen of an elderly woman in an assisted living facility in Minnesota. Smoke quickly poured out of the apartment and into a corridor that was shared with other residents because fire doors had been removed without authorization. The facility was protected with an automatic sprinkler system. The sprinklers in the kitchen and the entrance to the corridor activated and the fire was quickly extinguished. While there was minimal smoke and water damage, all residents were unharmed.

Automatic sprinkler systems are highly effective components of any building's fire protection plan. According to the U.S. Fire Administration (USFA), the presence of properly designed, installed, and maintained sprinkler systems reduces the chances of death from fire by one-third to two-thirds and property losses are cut in one-half or more.

Automatic sprinkler systems have been in use for well over a century. Their origin dates back to the Industrial Revolution when large mills were in operation in the northeastern part of the United States. The first sprinkler system was invented in 1874 by Henry S. Parmalee, who was looking for a way to protect his large piano mill.

The first standard for the design, installation, testing, and inspection of automatic sprinkler systems was written by the NFPA® in 1896. That standard, NFPA® 13, *Standard for the Installation of Sprinkler Systems*, has been in continuous publication since that date. There are currently two additional NFPA® standards that regulate the installation of automatic sprinkler systems: NFPA® 13D, *Standard for the Installation of Sprinkler Systems in One- and Two-Family Dwellings and Manufactured Homes*, and NFPA® 13R, *Standard for the Installation of Sprinkler Systems in Residential Occupancies Up to and Including Four Stories in Height*. All of these documents have been revised numerous times to reflect new technology, knowledge, and actual loss experiences.

Automatic Sprinkler System — System of water pipes, discharge nozzles, and control valves designed to activate during fires by automatically discharging enough water to control or extinguish a fire.

Today, automatic sprinkler systems are unsurpassed in fire protection. According to recent performance statistics published by the NFPA® (2008), when operated, **sprinklers** were effective 97 percent of the time. Failure to control the fires in sprinkler-equipped buildings were reportedly due to the following reasons:

- Improper maintenance

- Inadequate or inoperative water supply

- Incorrect design for the current hazard

- Obstructions **(Figure 6.1)**

- Partial protection

The purpose of this chapter is to address the various components and types of automatic sprinkler systems. Aspects of sprinklers and specialty sprinklers are also addressed. Procedures for the inspection and testing of sprinkler systems are highlighted as well. The information in this chapter is intended to be informative and descriptive, but should not be used as the authority over local codes, ordinances, and standards.

While there are a few commonly used agents in automatic fire sprinkler systems, the most common agent used is water. This chapter specifically addresses water-based automatic sprinkler systems. Suppression systems using other agents are addressed in Chapter 8, Special Extinguishing Systems.

Purpose of Automatic Sprinkler Systems

Modern automatic sprinkler system technology is both highly effective and reliable when properly designed, installed, and maintained. Today, sprinkler systems are installed in all types of structures. Many jurisdictions now

Figure 6.1 Obstructions, such as this ceiling fan, can disrupt the spray pattern from a sprinkler and can lessen its effectiveness. *Courtesy of Texas Fire Protection Specialists.*

require the installation of automatic sprinkler systems in one- and two-family dwellings. In fact, several U.S. communities have been requiring sprinkler systems in dwellings for several decades.

Fires that involve large unsprinklered properties pose a threat to the entire community and place an undue burden on fire-fighting resources. Reasons that building owners invest in and install automatic sprinklers include the following:

• Code requirements or by variance

• Potential incentives or reductions in insurance rates

• Insurance requirements

• General protection of life and property from fire

• Building design flexibility

• Inherent risk

Model building and fire codes require the installation of automatic sprinklers. The reasons for mandating sprinklers in buildings arise from a need to protect the community as a whole or to protect the occupants in individual buildings of high-occupancy design. Model codes require the installation of automatic sprinklers in buildings based upon their occupancy, construction type, and size **(Figure 6.2)**. When a building exceeds a given size established in a building code, it is typically required to have sprinklers. Good risk management practices suggest that automatic sprinklers be installed to protect the substantial investment in the structure and business even if the code does not require it.

Figure 6.2 Multistory hotels and other occupancies are typically required by building codes to incorporate automatic sprinkler systems.

Automatic sprinkler systems are also important to the life safety of firefighters. Controlling a fire in its earlier stages allows for a more **tenable atmosphere** for fire-fighting operations. This will increase the safety of the environment for firefighters to conduct suppression activities. Therefore, it is important that fire-fighting personnel understand how automatic sprinkler systems operate so that operations can be conducted more efficiently in buildings protected by these devices.

Tenable Atmosphere — Capable of maintaining human life.

The ideal fire control system should be simple, reliable, and automatic. It should use a readily available and inexpensive extinguishing agent and discharge the extinguishing agent directly on the fire while in its incipient stage. An automatic sprinkler system most closely meets these requirements.

Components of Automatic Sprinkler Systems

An automatic sprinkler system is an integrated system of pipes, sprinklers, and control valves designed to activate during fires by automatically discharging enough water to control or extinguish the fire **(Figure 6.3, p. 142)**. The system consists of a series of sprinklers that are systematically arranged so that the system will distribute sufficient quantities of water to either extinguish a fire or prevent its spread until firefighters arrive **(Figure 6.4, p. 142)**. Water in these systems is supplied to the sprinklers through a series of pipes.

Figure 6.3 Automatic sprinkler system designs are unique to each occupancy. *Courtesy of the Sand Springs (OK) Fire Department.*

Figure 6.4 A typical sprinkler.

Thermal Element — Device used in sprinklers and some fire detection equipment that is designed to activate when temperatures reach a predetermined level.

Deluge Sprinkler System — Fire-suppression system that consists of piping and open sprinklers. A fire detection system is used to activate the water or foam control valve. When the system activates, the extinguishing agent expels from all sprinkler heads in the designated area.

Preaction Sprinkler System — Fire-suppression system that consists of closed sprinkler heads attached to a piping system that contains air under pressure and a secondary detection system; both must operate before the extinguishing agent is released into the system.

Sprinklers are kept closed by **thermal elements**. When these devices are used, the heat from a fire causes those sprinklers above the fire to activate automatically. Some systems, such as **deluge**- or **preaction**-type sprinkler systems, operate when an electronic detector or manual control device is activated. An automatic sprinkler system typically consists of the following components:

- Suitable water supply
- Distribution piping
- Valves
- Sprinklers
- Fire Department Connection (FDC)

Sprinklers are designed to provide protection in a wide variety of situations. Although they may be fairly sophisticated in actual application, the fundamental concept is still simple. It is the simplicity of the automatic sprinkler system that gives rise to its greatest virtue — reliability. The individual components of an automatic sprinkler system are discussed in the sections that follow.

Suitable Water Supply

Every automatic sprinkler system must have a water supply of adequate volume, pressure, and reliability. Minimum water flow requirements for the system are determined by the hazard being protected, the occupancy classification, and the fuel-loading conditions. The water supply system must be able to deliver the required volume and pressure of water to the highest or most remote sprinkler in a structure while maintaining a minimum residual, or remaining, pressure in the system. Systems must have a primary water supply and may be required to have a secondary water supply.

The primary water supply may come from a public or private source. This type of connection is often the only water supply available. A private water supply will be necessary if no public supply is available. Private water supplies may originate from impounded water sources such as on-site ponds, reservoirs, wells, and storage tanks (**Figure 6.5**).

Storage tanks may be used as secondary sources and in some circumstances, such as residential systems, they may also be the primary source. Secondary sources of water supply also include large static water sources. For more information on water supply, refer to Chapter 4, Water Supply Systems.

Distribution Piping

Four basic pipe materials are used in sprinkler systems, including the following:

- Ferrous metal (also known as black steel)
- Copper
- Galvanized steel
- Plastic

The types of piping used in a sprinkler system are specified in the various NFPA® sprinkler system standards. Ferrous metal piping is the most common type in use (**Figure 6.6**). Ferrous metal piping has a long life expectancy but will eventually corrode. In addition, it is very heavy. This type of pipe is typically joined by threading, grooves, flanges, or welding.

The wall thickness of pipe is normally referred to as the **pipe's schedule**. The pipe's schedule directly influences water flow characteristics. Sprinkler piping is typically Schedule 10, Schedule 40, or special listed pipe.

Pipe Schedule — Thickness of the wall of a pipe.

Copper piping has been approved for sprinkler systems since the early 1960s (**Figure 6.7, p.144**). It is joined together by soldering or brazing. Copper piping is highly resistant to corrosion, has low friction loss, is lighter in weight than steel pipe, and is neat in appearance where exposed. It is rather expensive, however, and its use has declined with the increase of less-expensive plastic pipe.

Plastic pipe is the least expensive type of sprinkler system piping. If plastic pipe is used, it must be listed for use in sprinkler systems (**Figure 6.8, p. 144**). It is also the lightest in weight and the easiest to install. Plastic pipe is joined

Figure 6.5 Storage tanks are a key component of a private water supply.

Figure 6.6 Ferrous metal piping is the most commonly used type in automatic sprinkler systems.

Figure 6.7 While copper pipe is expensive, it does give an aesthetic detail to a structure when left exposed.

with plastic cement that is listed for that purpose. The main drawback of plastic pipe is that it may not be installed in areas where the ambient temperature exceeds specific limitations.

The various types and sizes of pipes in a sprinkler system are given specific functional names based on the role each serves within the system **(Figure 6.9)**. These include the following:

- *Water Supply Main* — Piping that connects the sprinkler system to the main water supply. The water supply main is the underground municipal main that is used to supply other systems such as fire hydrants and fire pumps. The water supply main may also be the private fire main that connects the municipal main to the sprinkler system.

Figure 6.8 Plastic pipe used to supply sprinklers must be listed for that use.

Figure 6.9 A typical sprinkler system with key components highlighted.

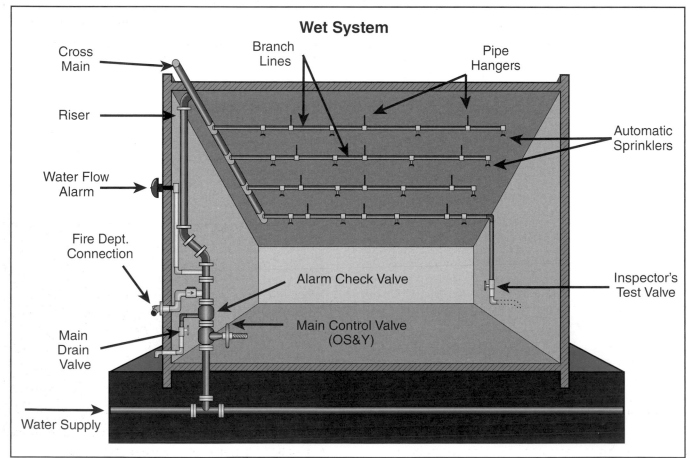

- *System Riser* — Vertical piping that extends upward from the water supply to feed the cross or feed mains. The system **riser** will typically have a control valve unless an outside control valve such as a post indicator valve (PIV) is used. The riser will have a drain, pressure gauge, and a waterflow alarm device attached to it.

- *Sprig* — Pipe that rises vertically and supplies a single sprinkler.

- *Riser* — Any vertical supply piping in the system.

- *Feed Main* — Pipe that supplies water to each of the cross mains.

- *Cross Main* — Pipe that feeds the branch lines.

- *Branch Lines* — Pipes that contain the individual sprinkler devices and include grid lines on gridded systems.

> **Riser** — Vertical water pipe used to carry water for fire protection systems aboveground such as a standpipe riser or sprinkler riser.

Valves

Every sprinkler system is equipped with various water-control valves and operating valves. Valves may be used to shut off the water, drain the system, prevent recirculation of water, and serve other needs of the system **(Figure 6.10)**.

Control Valves

The main control valve is used to shut off the water supply to the system when it is necessary to replace sprinklers or to perform other maintenance. These valves are located between the sources of water supply and the sprinkler system. After maintenance has been performed, the control valves must always be returned to the open position. Several major property losses have occurred because the water supply to the sprinkler system was shut off when a fire occurred.

Figure 6.10 Numerous valves are incorporated into a sprinkler system to serve specific purposes.

To help ensure that valves are not inadvertently left in a closed position, all valves that control water to sprinkler systems are of the indicating type. These valves must be supervised by manual or electronic means. For example, valves must be chained and padlocked in the open position and/or electronically connected to the fire alarm. With an **indicating valve**, the position of the valve — open or closed — can be determined at a glance. However, not all water-control valves are indicating-type valves. Underground valves in water distribution systems are commonly nonindicating valves.

> **Indicating Valve** — Water main valve that visually shows the open or closed status of the valve.

There are three types of control valves: butterfly valves, gate valves, and ball valves. Many systems use an outside stem and yoke (OS&Y) valve. OS&Y valves are gate valves that have a yoke on the outside with a threaded stem that controls the opening and closing of the gate by turning a hand wheel **(Figure 6.11)**. The threaded portion of the stem is outside the yoke when the valve is open and inside the yoke when the valve is closed. The position of an OS&Y valve is easily seen from a distance as opposed to other types of indicating valves.

Figure 6.11 Many control valves are OS&Y type.

Other types of indicating gate valves are the post indicator valve (PIV) and the wall post indicator valve (WPIV). The PIV is used to control underground sprinkler valves and consists of a hollow metal post attached to the valve housing **(Figure 6.12)**. The valve stem is inside this post. Mounted on the stem is a movable target with the words *OPEN* or *SHUT* visible through a window depending on the position of the valve. The operating handle is fastened and normally locked to the post. When the valve is closed, the word *SHUT* appears through the window. The WPIV is similar to the PIV, except that it extends through the building wall with the target and valve-operating wheel on the outside of the building **(Figure 6.13)**. A variation of the PIV is the indicating butterfly valve. This valve has a paddle indicator or a pointer arrow that shows the position of the valve.

The operating mechanism of sprinkler system control valves may be either the gate or butterfly type. A gate-valve mechanism consists of a close-tolerance gate that slides across a waterway **(Figure 6.14)**. In a butterfly valve, a disc rotates 90 degrees inside the waterway. Butterfly valves are operated by a worm gear, which is turned by a handle or a handwheel **(Figure 6.15)**. The position of a butterfly valve is indicated by a pointer that points either to the word *OPEN* or the word *CLOSED* on the valve body or by a cross-view of the valve showing its open and closed position.

A ball valve may be used as a control valve as long as it cannot be closed in less than 5 seconds **(Figure 6.16)**. Closing any valve in less than 5 seconds is always ill-advised because it may cause *water hammer*, resulting in damage to the system components. Ball valves that are listed for use in a fire sprinkler system as a control valve have a handwheel. These valves are typically found as floor-control valves. The ball valve has an indicating device on the body of the valve indicating whether or not the valve is open.

Figure 6.12 A typical PIV.

Figure 6.13 Multiple WPIVs in use at a big-box store.

Figure 6.15 The butterfly valve rotates inside the waterway.

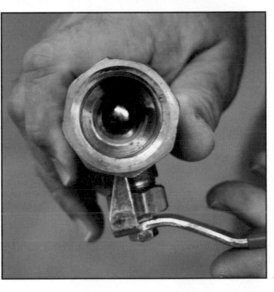

Figure 6.16 Ball valves must not be closed in less than 5 seconds in order to prevent water hammer.

Figure 6.14 To shut the valve, the gate valve slides down to close the waterway.

Operating Valves

In addition to control valves, sprinkler systems employ various operating valves such as alarm valves, check valves, automatic drain valves, globe valves, and stop or cock valves. An **alarm check valve** is specially designed to incorporate an alarm function. When one or more sprinklers open, some of the water is directed through a pipe that has a pressure switch and/or water motor gong that activates the alarm. Most systems with alarm check valves provide for a **retard chamber** to help prevent false alarms caused by water surges. The alarm check valve is also equipped with a test valve. This small valve diverts water from the water supply into the alarm pipe and retard chamber, which allows for the testing of alarms without tripping the alarm check valve. An alarm test valve is located on the riser and is used to flow water for the purpose of testing the water-flow alarm.

Check valves. These valves are used to limit the flow of water to one direction **(Figure 6.17, p. 148)**. They are placed in water sources to prevent recirculation or backflow of water from the sprinkler system into the municipal water

Alarm Check Valve — Type of check valve installed in the riser of an automatic sprinkler system that transmits a water-flow alarm when the water flow in the system lifts the valve clapper.

Retard Chamber — Chamber that that catches and slows the excess water that may be sent through the alarm valve of an automatic sprinkler system during momentary water pressure surges. This reduces the chance of false-alarm activation. The retard chamber is installed between the alarm check valve and alarm-signaling equipment.

supply. These valves are also required in the FDC line. **Check valves** usually have an arrow cast into the body to indicate the direction of flow. If there is no arrow or indication, the valve clapper pivot should be on the end toward the source of supply.

***Automatic drain valves.* Automatic drain valves** (also known as *ball drip valves*) drain piping when pressure is relieved in the pipe. These can be main drain valves or auxiliary drain valves. The most common application of these valves is to drain water from Siamese connections or the FDC after use in certain types of sprinkler systems. This prevents the portion of the pipe that extends through the wall from freezing in cold weather.

Globe valves. These are small handwheel-type valves that are primarily used on drains and test valves **(Figure 6.18)**. This includes the inspector's test valve, which is used to cause water to flow through the system for an inspection test. The inspector's test valve may be located near a riser, for convenience, or at a remote location within the sprinkler system.

***Stop* or *cock valves*.** These valves are also used for drains and alarm testing. Ball-type valves are most commonly opened and closed with a quarter turn of the valve.

Figure 6.17 Check valves limit the flow of water to one direction.

Sprinklers

The sprinkler is the portion of the sprinkler system that reacts to the heat of a fire and then delivers water to the fire area. A simple thermally-sensitive device controls most automatic sprinklers. Common types of these devices are the fusible link and frangible bulb **(Figures 6.19 a and b)**. In its simplest form, the *fusible link* is a solder link with a low, precisely established melting point. The solder link is connected to a cap that restrains the water at the orifice. When there is a fire, the solder in the fusible link melts at its predetermined melting point. The lever arm or struts are released and spring clear of the sprinkler frame **(Figure 6.20)**. As the lever arm drops, the seated cap is released, which permits the water to flow.

Frangible bulbs are inserted between the sprinkler frame and the discharge orifice in much the same way as the fusible link. Frangible bulbs contain a liquid that expands when heated. When the expansion of the liquid within the bulb exceeds the strength of the bulb, the bulb breaks, allowing water to flow.

Sprinklers can be modified for different environments. For example, if a sprinkler is to be placed in a corrosive atmosphere such as a plating room, it can be coated with wax for protection and listed for such use by the manufacturer. If mechanical damage is possible,

Figure 6.18 Globe valves are primarily used on drains and test valves.

Figure 6.19a A fusible link sprinkler.

Figure 6.19b A frangible bulb sprinkler.

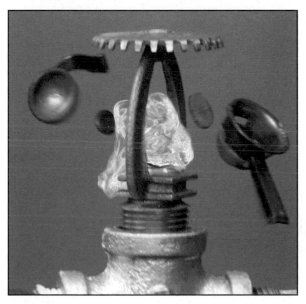

Figure 6.20 As the solder in the fusible link melts, the lever arms or struts are released.

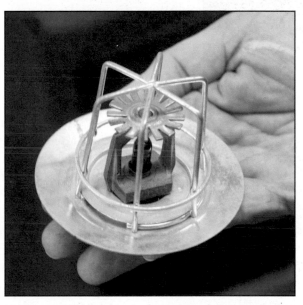

Figure 6.21 Sprinklers are often fitted with protective cages in areas such as gymnasiums where they may become damaged.

the sprinkler can be fitted with a protective cage **(Figure 6.21)**. In areas where appearance is important, such as in a residential setting, a variety of recessed, flush, and concealed sprinklers are available that have a finish matching the color and texture of the ceiling **(Figure 6.22)**. Such finishes must be applied by the sprinkler manufacturer, *not* the contractor or occupant.

Sprinklers have certain characteristics such as the various temperature ratings, response times, and deflector types. These will be discussed later in this chapter.

Figure 6.22 A typical recessed sprinkler.

Figure 6.23 Traditional FDCs have two 2½ inch (65mm) female inlets to which fire hoses can be connected.

Figure 6.24 Some new style FDCs incorporate inlets for large diameter hose (LDH).

Fire Department Connection (FDC)

Commercial and certain residential sprinkler systems are required to provide one or more fire department connections (sometimes called *Siamese connections* or simply the *FDC*). These connections permit the fire department pumper to pump into the sprinkler system, thereby boosting the pressure and the volume of water in the system. Fire department connections are especially important to the protection of a building in which the primary water supply is weak, the supply being overtaxed by a large number of open sprinklers, or where the system is old and heavy pipe corrosion is suspected.

The most common type of FDC connection is a 4-inch (100 mm) pipe equipped with a Siamese fitting on the outside of the building. The Siamese fitting has two 2½ inch (65 mm) female inlets to which fire hoses can be connected **(Figure 6.23)**. Larger systems may have three or more inlets. Some newer connections may be equipped with inlets for large diameter hose (LDH) **(Figure 6.24)**. Fire department connections and their requirements are discussed in more depth in Chapter 7, Standpipe and Hose Systems.

Important Aspects of Sprinklers

As was discussed earlier, sprinklers, sometimes referred to as *sprinkler heads* or *heads*, are the part of the suppression system that delivers water to the fire area. There are three main characteristics of a sprinkler that are of interest: temperature ratings, response times, and deflector components.

Temperature Rating

The selection of the operating temperature of a sprinkler is determined by the maximum air temperature expected at the level of the sprinkler under normal conditions. Operating temperatures vary from 135°F to 600°F (57°C to 315°C) or higher. For ordinary room temperatures, a sprinkler with a temperature rating of 135°F to 170°F (57°C to 77°C) is most frequently used. If the typical ambient temperature exceeds 100°F (38°C), such as in an attic or near a heater, a sprinkler with a higher temperature rating is used. It is necessary to provide a margin between normal room temperature and the operating temperature in order to prevent inadvertent activation.

The components of a sprinkler can be varied to fit specific applications. By changing the composition of the solder, the operating temperature of the sprinkler can be changed. It is also possible to vary the operating temperature of sprinklers using other types of thermally-sensitive elements.

Temperature ratings that are too low for a given location may result in malfunction or accidental activation. Temperature ratings that are too high will delay sprinkler operation, enhancing fire growth. Some insurance carriers recommend the use of high-temperature sprinklers in industrial and warehouse occupancies as a way to limit the total number of sprinklers likely to activate during a fire. These high-temperature sprinklers reduce water damage by concentrating water flow in the immediate fire area rather than having heads that are remote to the fire activate.

Sprinkler frames or glass bulbs are color-coded so that their temperature ratings can be distinguished quickly **(Figure 6.25)**. For example, those sprinklers rated for temperatures from 250°F to 300°F (121°C to 149°C) are blue. The temperature rating is also stamped on the link in fusible link-type sprinklers. On other types of sprinklers, the temperature rating is stamped on some other part of the sprinkler.

Figure 6.25 Both fusible link and frangible bulb sprinklers are color-coded so that temperature ratings can be quickly distinguished.

Sprinkler Response Time

Because most sprinklers are thermal devices, there is some delay between the ignition of a fire and the operation of the sprinkler. This delay is a function of several variables, including the design of the sprinkler.

The activation time of the sprinkler depends on the surface area, mass, and thermal characteristics of the heat-sensitive element. Although the gas temperature surrounding a sprinkler may be 165°F (74°C), a heat-sensitive element rated at 165°F (74°C) will not reach this temperature for some time, depending on its surface area, mass, and thermal characteristics. In addition, the temperature and velocity of a gas jet as it travels across the ceiling will affect the activation time because heat must be transferred from the hot gas jet to the operating elements of the sprinkler. The faster and hotter the gas jet, the faster the sprinkler will operate.

To speed the operation of sprinklers, engineers have designed types of sprinklers known as early-suppression fast-response (ESFR) sprinklers. These faster-operating sprinklers can be compared to ordinary sprinklers through the use of a Response Time Index (RTI). ESFR sprinklers typically react 5 to 10 times faster than traditional sprinklers. The lower a sprinkler's RTI, the faster it responds. ESFR sprinklers can usually be identified quickly because they are larger than conventional sprinklers.

Deflector Component

Another important component of a sprinkler is the **deflector**, which is attached to the sprinkler frame. The deflector creates the discharge pattern of the water **(Figure 6.26)**. Pressure forces water against the deflector to convert it into a spray pattern.

Deflector — Part of the sprinkler assembly that creates the discharge pattern of the water.

Sprinklers produced before 1955 (commonly known as *old-style sprinklers*) were designed with deflectors that discharged a large portion of the water upward toward the ceiling in order to protect structural elements. Unfortunately, this design did not produce a good downward distribution of water. Modern

Figure 6.26 The deflector is attached to the sprinkler frame and creates the discharge pattern of the water.

standard sprinklers produce a more uniform discharge pattern that is directed downward. By directing all of the water downward, the fire is controlled more effectively, resulting in reduced ceiling temperatures and better protection for structural elements. Because of the difference in discharge patterns, old-style sprinklers cannot be used to replace modern sprinklers. Modern sprinklers may be substituted for old-style sprinklers in an existing system if it becomes necessary to change or upgrade a sprinkler system.

The deflector configuration is fundamental to the effectiveness of the sprinkler. Basic sprinkler deflector types include the following:

- *Upright* — Designed to deflect the spray of water downward in a hemispherical pattern (**Figure 6.27**). Upright sprinklers cannot be inverted for use in the hanging or pendant position because the spray would be deflected toward the ceiling. These are typically used in dry systems.

- *Pendant* — Used where it is impractical or unsightly to use sprinklers in an upright position such as below a suspended ceiling **(Figure 6.28)**. The deflector on this type of sprinkler breaks the pattern of water into a circular pattern of small water droplets and directs the water downward.

- *Sidewall* — Used in instances where it may be desirable or required to install sprinklers on the wall at the side of a room or space **(Figure 6.29)**. This may be for cost-savings or appearance. By modifying the deflector, a sprinkler can be made to discharge most of its water to one side. Sidewall sprinklers are

Figure 6.27 A typical upright sprinkler.

Figure 6.29 Sidewall sprinklers are commonly found in corridors and hotel rooms.

Figure 6.28 Pendant sprinklers are used in applications where an upright sprinkler is not practical, such as below a suspended ceiling.

useful in areas such as corridors, offices, hotel rooms, and residential occupancies. For a variety of reasons, extended-coverage sidewall sprinklers have become more popular in certain occupancies. One reason their popularity has increased is that an extended-coverage sidewall sprinkler can be positioned at one end of a rectangular room, thereby reducing piping costs.

- *Concealed* — Hidden by a removable decorative cover that releases when exposed to a specific level of heat (**Figure 6.30**).

- *Flush* — Mounted in a ceiling with the body of the sprinkler, including the threaded shank, above the plane of the ceiling.

- *Recessed* — Installed in recessed housing within the ceiling of a compartment or space; all or part of the sprinkler other than the threaded shank is mounted in the housing.

- *In-Rack* — Typically used in storage facilities. In-rack sprinklers incorporate a protective disc that shields the heat-sensing element from water that is discharged from the sprinklers above **(Figure 6.31)**.

Figure 6.30 A typical concealed sprinkler assembly.

Figure 6.31 The disc attached to an in-rack sprinkler protects the heat-sensing element from water that is discharged from sprinklers above.

Specialty Sprinklers

While the types of sprinklers previously listed can be used in a variety of circumstances, there are situations where a special type of sprinkler may be best suited or required by the codes adopted by the authority having jurisdiction (AHJ). These might be situations that call for a particular pattern of water distribution or unusual ambient temperatures that mandate special construction. Common types of specialty sprinklers include the following:

- *Large drop* — Designed to produce large drops of water to enable the spray to penetrate strong updrafts created by elevated fire conditions.

- *Dry pendant* — Used in either dry or wet systems where the protected area may freeze. An example would be walk-in freezer in a store or storage facility.

- *Residential* — Listed for use in residential occupancies. These are fast-response sprinklers that have a special low-mass fusible link or bulb that makes the time of temperature actuation much less than that of a conventional sprinkler.

- *Water spray nozzle* — Discharges water in a specific pattern to protect a three-dimensional hazard such as an electric transformer or flammable liquid or gas tank **(Figure 6.32)**. These sprinklers may or may not have a deflector and must be installed in a specific orientation to properly wet the surface of the hazard.

Figure 6.32 Water spray nozzles protect three-dimensional hazards such as electric transformers.

Figure 6.33 Attic sprinklers are designed to provide protection to roof structures.

- *Attic* — Designed to protect sloped attic spaces. Their unique discharge patterns were developed to produce a narrow but long pattern **(Figure 6.33)**. One of their primary functions is to provide protection of the roof structure.

- *Water mist* — Turns the water into a very fine mist, which depletes oxygen and blocks radiant heat. There are several different water-mist sprinkler designs available that can be matched for the specific hazard or application. Water mist sprinkler systems and their application are discussed further in Chapter 8, Special Extinguishing Systems.

Types of Sprinkler Systems

While the components of all sprinkler systems are basically the same, there are different types of sprinkler systems depending upon the needs and requirements for the occupancy and environment. The most common types of sprinkler systems in use today include the following:

- Wet-pipe systems
- Dry-pipe systems
- Deluge systems
- Preaction systems
- Residential systems
- Storage facility systems

Wet-Pipe Systems

Wet-Pipe Sprinkler System — Fire-suppression system that is built into a structure or site; piping contains either water or foam solution continuously; activation of a sprinkler causes the extinguishing agent to flow from the open sprinkler.

A **wet-pipe sprinkler system** contains water under pressure in the piping at all times so that the opening of a sprinkler immediately discharges water onto a fire and activates an alarm **(Figure 6.34)**. Wet-pipe sprinkler systems are popular due to their simplicity and reliability, and are usually successful in controlling fires with only a minimum of sprinklers opening. Research by the NFPA® indicates that 99 percent of fires are controlled with less than 10 sprinklers activated.

Figure 6.34 Wet-pipe sprinkler systems contain water under pressure in the piping at all times.

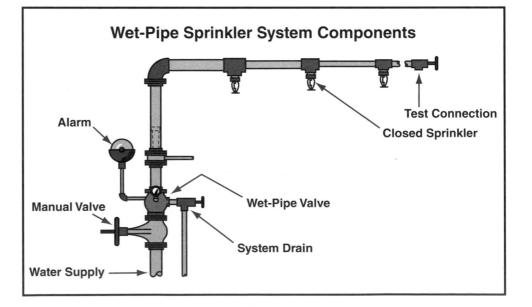

Wet-Pipe Sprinkler System Components

Alarm · Test Connection · Closed Sprinkler · Manual Valve · Wet-Pipe Valve · System Drain · Water Supply

Depending on the size of the systems and the occupancies in which they are built, wet-pipe systems fall under the requirements of NFPA® 13, 13D, or 13R. Standard wet-pipe systems may be installed in any location where system components will not be subject to freezing temperatures. Wet-pipe systems of limited size may be installed in areas subject to freezing if they are properly freeze-protected.

These systems are typically equipped with a **water-flow indicator** that initiates an alarm when water begins to flow in the system. A water-flow switch consists of a vane or paddle that protrudes through the riser into the waterway **(Figure 6.35)**. The vane is connected to an alarm switch on the outside of the riser. Movement of the vane, caused by flowing water, operates the switch that initiates an alarm. The vane must be thin and pliable so that if many sprinklers operate, the water flow will flatten the vane against the wall of the riser, resulting in a clear waterway.

Just as the retard chamber helps to prevent false alarms with an alarm check valve, there is a time-delay feature on water-flow switches. This delay is typically 30 seconds but may be up to 90 seconds from activation depending on AHJ requirements. Water-flow switches are compact and used frequently where only an electrical output from an alarm is desired.

As was mentioned earlier, when a wet-pipe sprinkler system must protect a small, unheated area, such as a loading dock or residential balcony, an antifreeze system may be necessary. In an antifreeze system, the ordinary water

Figure 6.35 Water-flow indicators initiate an alarm when water movement is detected in the system.

Figure 6.36 A U-loop piping arrangement prevents the antifreeze solution from leaking back into the rest of the system.

in the sprinkler piping is replaced with a non-freezing, nonflammable liquid. Increased concern for environmental health has resulted in regulations restricting the type of chemical that can be used in any piping system connected to a public water supply. However, certain grades of glycerin or propylene glycol are usually acceptable. In systems not connected to a public water supply, other liquids can be used. The cost and required annual maintenance of these antifreeze liquids usually restricts their use to systems of less than 40 gallons (150 L).

Antifreeze systems are constructed in the shape of a U-loop **(Figure 6.36)**. This arrangement prevents dilution of the antifreeze solution and keeps the antifreeze out of the main portion of the system. Reduced pressure zone backflows may be required to ensure nothing from the antifreeze system reaches the potable water supply. In addition, a listed expansion tank may be required.

Dry-Pipe Systems

Dry-Pipe Sprinkler System — Fire-suppression system that consists of closed sprinklers attached to a piping system that contains air under pressure. When a sprinkler activates, air is released that activates the water or foam control valve and fills the piping with extinguishing agent.

A **dry-pipe sprinkler system** is one in which air or nitrogen under pressure replaces water in the pipes. When buildings lack sufficient heat to keep the water in the sprinkler pipes from freezing, a dry-pipe sprinkler system must be used. The dry-pipe valve must be located in a heated portion of the structure, which must be maintained at a minimum of 40°F (4°C) according to code requirements. If the valve is subject to freezing conditions, it may not activate properly when required. In some cases, it may be necessary to build a small, heated enclosure specifically for the dry-pipe valve riser and its associated equipment.

For a dry-pipe system to operate, heat from a fire must cause the sprinkler to activate. Pressurized air contained in the piping will flow through the open sprinkler. After a slight drop in air pressure, the quick-opening device (if present) will activate to accelerate the removal of air from the system. Once the air pressure is significantly reduced, the dry-pipe valve trips open. Water will then enter the intermediate chamber of the dry-pipe valve and flow through the entire piping system. It will then be discharged through the open sprinkler.

The air necessary to service a dry system may be obtained from two sources. One source is an air compressor and tank arrangement dedicated for exclusive use with the system. The second alternative is to have the sprinkler system connected to a plant or shop air system from a reliable source.

A device known as a *dry-pipe valve* uses the force created by the air pressure in the system to hold the clapper in the valve closed and keep water from entering the system until a sprinkler operates **(Figure 6.37)**. When a sprinkler opens, there is a reduction in the system air pressure and the dry-pipe valve opens. This admits water into the sprinkler system.

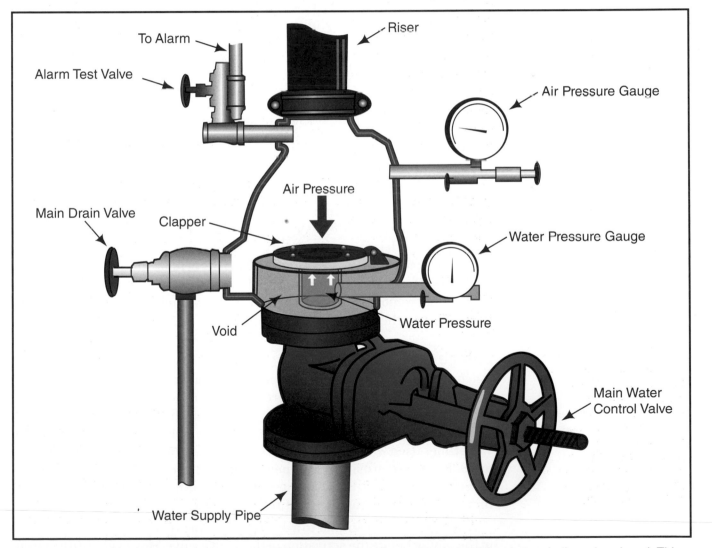

Figure 6.37 Dry-pipe valves use the force created by air pressure in the system to hold the clapper in the valve closed. This keeps water out of the system until a sprinkler is activated.

The dry-pipe valve uses a differential principle incorporated into the valve design to open. When the differential dry-pipe valve operates and the clapper swings open, the valve is held in place by a latch to prevent it from reseating. Releasing the latch requires that the valve body be opened. Therefore, these systems must be manually reset after they have activated.

The air-water differential allows a small amount of air pressure to hold back the water. This differential is a ratio of air pressure to water supply pressure within the system to maintain the dry-pipe valve in the closed system. The differential is generally a 3:1 to 7:1 ratio where, for example, 1 psi (7 kPa) of air is needed to hold back 3 psi (20 kPa) of water. In addition, the less air pressure in the system means the quicker water can expel the air and therefore discharge water on the fire. The air pressure in the system must be maintained within the prescribed limits, which are typically no more than 20 psi (140 kPa) above the pressure at which the dry-pipe valve will trip.

The ordinary dry-pipe valve, which is widely used, has a couple of shortcomings. First, it is slow to operate. Second, when the valve trips, the water rushes into the system with considerable velocity. The inrush of water can cause damage to the system's components. This can be corrected with the use of low-differential dry-pipe valves. However, low-differential dry-pipe valves require the use of a heavy-duty air compressor and an automatic pressure-maintenance device.

Dry-pipe systems can use quick-opening devices to help speed up the process of getting water to the source of the fire. In large dry-pipe systems, several minutes can elapse as air is being expelled through the open sprinklers before water is discharged. Quick-opening devices limit this delay. There are two types of quick-opening devices: accelerators and exhausters. An *accelerator* works by unbalancing the differential in the dry-pipe valve, which causes it to trip more quickly. The accelerator detects a drop in system pressure and forces the clapper open to allow water into the system. An *exhauster* functions by quickly expelling air from the system. The exhauster is generally located at the most remote end of the system from the water supply and does not work as fast as the accelerator. Quick-opening devices are required on any dry-pipe valve that serves a piping system with a capacity of more than 500 gallons (1 900 L). If the system capacity is over 750 gallons (2 800 L), water must be provided to the most remote sprinkler within 15 to 60 seconds of system activation, depending on the occupancy classification.

Deluge Systems

A *deluge system* is designed to quickly supply a large volume of water to the protected area. Deluge systems are normally used to protect extra-hazard occupancies such as aircraft hangars, woodworking shops, cooling towers, ammunition storage, or certain manufacturing facilities. Deluge systems are sometimes used with foam to protect against flammable liquid hazards (**Figure 6.38**). A system using some open and some closed sprinklers is a variation of the deluge system. Because these systems require large volumes of water, they are sometimes supplied by fire pumps.

A deluge system is similar to a dry-pipe system in that no water is contained in the piping before the activation of the deluge valve. However, it differs from a dry-pipe system in that all of the sprinklers are open, with no fusible links.

Figure 6.38 Areas such as aircraft hangars are often protected by deluge foam systems.

This means that when the valve is tripped and water enters the system, the water will discharge through all the sprinklers simultaneously. The flow of water to the system is controlled by a deluge valve. Fire detection devices, which may be heat, smoke, and/or flame detectors, are installed in the same area as the sprinklers. The detection devices control the operation of the deluge valve through a tripping device and are required to be automatically supervised. Unlike wet- or dry-pipe systems, the sprinklers do not function as the detection device in a deluge system.

A **deluge valve** keeps water out of the pipes until the system is activated. Just as there are several types of fire protection systems, there are also several methods of operating the deluge valve. They can be operated electrically, pneumatically, or hydraulically. Deluge valves also have provisions for manual operation. Some operations are described as follows:

- *Electrical operation* — An electrically operated deluge valve is designed for use with fire detectors that transmit an electrical signal to the valve. The activation system includes a releasing solenoid that opens the deluge valve clapper. Also included is a manual release and reset in case of power failure. Normal electrical service is a primary power source; however, batteries are usually provided as a secondary power source. A major advantage of the electrically operated deluge valve is its speed of operation. Most modern deluge systems are activated electronically and use similar components to a fire alarm system.

Deluge Valve — Automatic valve used to control water to a deluge sprinkler system.

- *Pneumatic operation* — Pneumatically operated deluge valves are designed for use in a pneumatic detection system, usually one that has rate-of-rise detectors. This type of activation is actually a combination of pneumatic and mechanical activation. A pneumatic detector operates on an imbalance of pressure. The deluge valve activates when a change in air pressure within the device (caused by an increase in temperature) displaces a diaphragm. The diaphragm is mechanically linked to a tripping mechanism that in turn releases the deluge valve clapper. The pneumatic system does not require electrical power for activation, but a manual release is required.

- *Hydraulic operation* — There are several different types of hydraulically operated deluge valves. Some of these valves may use a combination of hydraulically operated valves and a pneumatic, hydraulic, or electrical detection system. Regardless of the detection system used, hydraulic pressure is usually required for activation by changing the differential pressure in the deluge valve.

When a heat, smoke, and/or flame detector senses the presence of a fire condition, the fire detection system sends a signal to the deluge valve, causing the valve to open. In a situation where the fire is discovered by an individual, the deluge valve can be manually activated. As water enters the deluge valve and the piping, a pressure switch is activated that transmits an alarm either locally or to a supervising station. If present, a water motor gong is activated. The water then flows through all open sprinklers simultaneously.

Preaction Systems

A *preaction system* is used when it is especially important that the inadvertent release of water be minimized, even if the sprinkler pipes accidentally break. Preaction systems are frequently used to protect computer rooms, document-storage areas, freezers, or cold-storage warehouses. The system employs a deluge-type valve, fire detection devices, and closed sprinklers. Like deluge systems, the preaction system valve will not discharge water into the system's piping until an indication is received from fire detection devices that a fire may exist. Once water is in the system, it may then be discharged through any sprinkler that has opened.

The piping in a preaction system is normally dry. Therefore, it can be used in freezing environments such as cold-storage warehouses. Where the preaction system exceeds 20 sprinklers, regulations require that the piping be supervised. This is usually accomplished with air under low pressure. In the event of a leak or break in the piping, the air pressure drops and transmits an alarm without admitting water into the system. Three basic categories of preaction sprinkler systems are as follows:

- *Noninterlock system* — Admits water to sprinkler piping upon operation of detection devices or automatic sprinklers.

- *Single interlock system* — Admits water to the sprinkler piping upon operation of a detection device.

- *Double interlock system* — Admits water to sprinkler piping upon operation of both detection devices and automatic sprinklers or two separate detection devices or circuits.

When a detector senses the presence of a fire condition, a signal is sent to the preaction valve, causing the valve to open. Sensors in the piping system detect the flow of water into the system and trigger the waterflow fire alarm. When the level of heat at a sprinkler reaches the appropriate temperature, the sprinkler opens and water flows through the orifice.

Residential Systems

A residential sprinkler system is an automatic sprinkler system that is specifically designed to enhance the survivability of individuals that are in the room of fire origin **(Figure 6.39)**. The primary purpose of these types of systems is to reduce injury or loss of life due to a fire. An added benefit is the potential reduction of property damage. Residential systems are expected to prevent flashover in the room of fire origin and improve the chance for the occupants to escape or be evacuated.

Figure 6.39 A typical residential sprinkler head.

Residential systems designed for one- and two-family dwellings are smaller and more economical than those for commercial occupancies. The water supply source is generally the same as the domestic water supply. This approach works because the water supply requirements are substantially less for residential systems. A minimum of a 10-minute supply of stored water is required for systems designed in accordance with NFPA® 13D. If structures are less than 2,000 square feet (185 m²) in area and no more than one story in height, the minimum quantity is based on the flow rate for two sprinklers times 7 minutes of operation. In residential structures such as hotels and motels that are 4 stories or less, NFPA® 13R requires a minimum of a 30-minute supply.

To make sprinkler systems useful in residential application, there are a few changes in design, operation, water supply, and flow requirements. These changes decrease the cost of the system while enhancing its effectiveness in protecting life and property. Some of these changes include the following:

- Modification of sprinkler design and the development of fast-response residential sprinklers.

- Minimum flow requirements of 18 gpm (68 L/min) from an individual sprinkler for residential protection.

- Alarms that are simpler and better designed for residential applications.

A major difference between residential sprinklers and standard sprinklers is their sensitivity or speed of operation. Residential sprinklers operate more quickly than standard sprinklers. By redesigning the fusible link, even at 165°F (74°C), the sprinkler can be made to operate before conditions in the room become so untenable that occupants cannot survive.

Sprinkler coverage in residential systems is not as extensive as in standard commercial systems. Sprinklers can be omitted from areas such as garages, carports, closets, and small bathrooms, for example. However, local codes may be amended to include some of these traditionally exempted areas. Residential sprinklers are designed to discharge water higher on the walls of a room to prevent a fire from traveling above the spray, which might occur with burning drapes or in preflashover conditions.

Figure 6.40 Residential sprinkler systems are often installed with piping materials not found in commercial installations.

These systems are installed with a variety of piping materials not generally found in commercial installations **(Figure 6.40)**. In addition, piping methods such as multipurpose or combination domestic and sprinkler water lines are used to supply both the domestic needs of a residence and the fire sprinklers.

To be of value, a residential sprinkler system must continually be in service. As with a standard system, inadvertent or deliberate closing of valves renders the system useless. Therefore, using one valve to control both the sprinklers and the water service for the residence eliminates the possibility of the supply valve being turned off. The sprinklers cannot be turned off without the household water supply being turned off as well.

The water supply for residential sprinklers may be taken from several sources. This can include a connection to the public water system, an on-site pressure tank, or a storage tank with an automatic pump. A connection to the public water supply is reliable and will usually provide adequate volume. Public water systems may not service rural homes, which would require the use of a pressure tank or tank with an automatic pump. Large homes or homes found in urban areas may also need a fire pump and/or tank based on hydraulic calculations.

Storage Facility Systems

Storage facilities present a particular set of fire risk issues. If storage space is used efficiently, the burning rate and fire spread is increased should an unwanted fire occur. The higher loss potential can be attributed to large undivided areas, large concentrations of value in a single fire area, and very high stockpiling with narrow aisles **(Figure 6.41)**. To protect these areas effectively, firefighters must have knowledge of the sprinkler system and the storage methods associated with the warehouse. NFPA® standards outline methods for providing fire protection for storage facilities based upon commodity classification and the storage arrangement.

Identifying the commodities that are stored and the method of their storage are essential tasks for analyzing the amount of sprinkler protection needed for high-storage areas. There are several major categories of commodity classification for storage facilities. These classifications reflect the burning rate and heat release rate of the commodity as well as the effect that water has on the commodity. Commodity classifications are as follows:

- *Class I* — Generally noncombustible and stored on wood pallets in ordinary packaging. Class I commodities can be packaged in corrugated cardboard or stretch-wrapped as a unit load.

- *Class II* — Noncombustible commodities but packaged in wooden crates or multilayered cardboard cartons.

- *Class III* — Combustible materials such as wood, paper, or certain plastics, regardless of packaging.

Figure 6.41 Storage areas that stock materials high with narrow aisles can have a high loss potential.

Table 6.1 Types of Plastics		
Fastest Burning Group A	Moderate Burning Group B	Slowest Burning Group C
Acrylic	Cellulose Acetate	Melamine
Polycarbonate	Ethyl Cellulose	Phenolic
Polyethylene	Nylon	Polyvinyl Fluoride
Polystyrene	Silicone Rubber	Urea Formaldehyde

- *Class IV* — Class I, II, or III products that contain limited amounts of *Group A plastics*. Plastics present a special fire-control problem because they produce more heat per unit of weight than ordinary combustibles. Plastics are divided into three groups based upon their heat per unit of weight **(Table 6.1)**.

The classification is based upon the most severe hazard in the storage area. Miscellaneous types of storage have commodities of several classifications. It is better to assign a higher classification than originally contemplated because of the possibility of higher-classed commodities being introduced later. This might include a small increase in cost at installation but is typically cheaper than upgrading the system once it has been installed. For more information on commodity classifications, refer to NFPA® 13.

Testing and Inspection

Sprinkler systems require periodic inspections and maintenance in order to perform properly during a fire situation. Company managers, maintenance personnel, and fire prevention and inspection personnel should be able to inspect systems and identify problems **(Figure 6.42, p. 164)**. In addition, model codes, AHJs, and insurance companies may require periodic inspections to ensure proper operation and maintenance of the system. Ultimately, the building owner is responsible for inspection, testing, and maintenance of

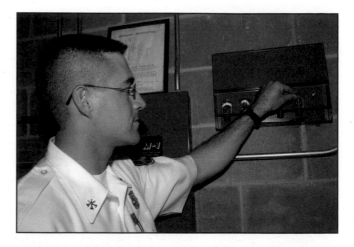

Figure 6.42 Inspections should be done as required by the AHJ to ensure the sprinkler system is operational and properly maintained.

the fire protection systems. In complex situations, or those where the building owner may be absent or off-site, those responsibilities may be transferred to another qualified person or company.

NFPA® 25, *Standard for the Inspection, Testing, and Maintenance of Water-Based Fire Protection Systems*, provides the requirements for periodic testing, maintenance, and inspection of automatic sprinkler systems. The following information introduces inspection and testing procedures but is not an all-inclusive review of the inspection process. For more information, please consult IFSTA's *Fire Inspection and Code Enforcement* manual, 7th Edition.

Acceptance Tests

In many jurisdictions, a representative from the AHJ or fire department may be required to witness sprinkler system acceptance tests. A representative of the installation contractor should conduct these tests. This releases the fire department from liability resulting from damaged equipment because of improper operation or installation. Depending on the type of system, the tests and procedures performed may include the following:

- Flushing of underground connections
- Hydrostatic tests (underground and overhead piping)
- Water-flow alarm test
- Main drain test
- Trip test (for dry-pipe systems)

IFSTA's *Fire Inspection and Code Enforcement* manual contains more information on acceptance testing and a variety of forms that may be used for sprinkler system inspections and testing.

Sprinklers

Sprinklers need to be inspected to ensure that they are appropriate for the occupancy and in good working order. The inspection should verify that all sprinklers are clean and not painted, undamaged, and free of corrosion **(Figure 6.43)**. It should also be noted whether guards are needed to protect sprinklers against mechanical damage.

Weak sprinklers can be detected by a noticeable change in the position of the fusible link or leaking around the sprinkler orifice. Sprinklers exposed to a corrosive atmosphere should have a special protective coating. Sprinklers that are corroded, painted, or loaded with foreign materials should be replaced. When inspecting the sprinklers themselves, it should be noted that a clearance of at least 18 inches (450 mm) should be maintained under sprinklers **(Figure 6.44)**. Some sprinklers, such as large drop and ESFR sprinklers, require a minimum clearance of 36 inches (900 mm). For this reason it is important to review the code requirements for the individual occupancy. Facilities should have a supply of extra sprinklers available so that they can be replaced when a fire occurs or when an inspection determines that repairs are needed **(Figure 6.45)**.

Figure 6.43 Sprinklers are often encountered that have been corroded.

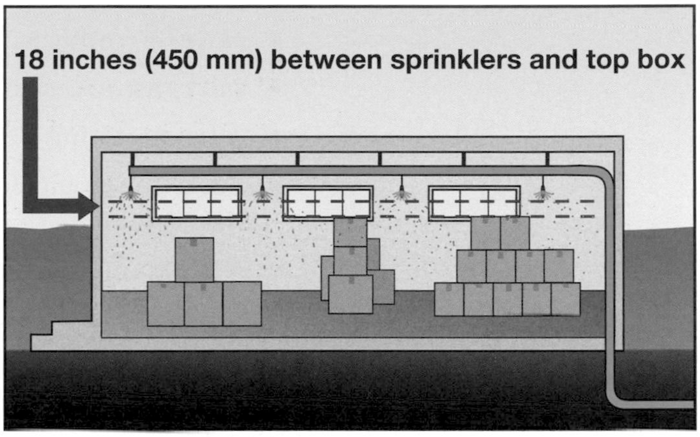

18 inches (450 mm) between sprinklers and top box

Figure 6.44 There should be at least 18 inches (450 mm) between sprinklers and stored materials.

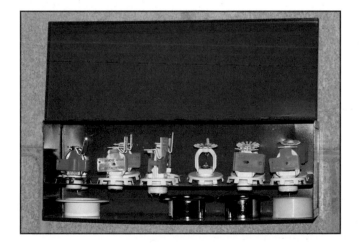

Figure 6.45 Each protected premises should have a supply of extra sprinklers available should replacement be necessary.

Testing and Inspection of Sprinkler Systems

All systems, regardless of type or occupancy, should be visually inspected on a monthly basis. Weekly or even daily inspections are recommended depending upon certain conditions such as cold weather. The detection systems should be tested semiannually and the system itself should be trip-tested annually. A written record of all tests and inspections should be made, compared to earlier tests, and kept on file for future reference.

When testing is complete, the alarm-monitoring organization (if one is used) should confirm that the alarm equipment functions properly. If no alarms were received (water-flow, valve supervision, etc.), corrective maintenance service is required on the alarm system. This should be coordinated between both alarm and sprinkler contractors.

Wet-Pipe Systems

Inspections of wet-pipe systems are primarily concentrated in four areas: valves, sprinklers, piping, and water supply. Inspectors should ensure that the following items are in operational condition:

- Valves are fully opened and secured or otherwise supervised in an approved manner.

- Valve operating wheels or cranks are in good condition **(Figure 6.46)**.

- Valves are accessible at all times.

- Valve operating stems have not been subjected to mechanical damage.

- Operating wrenches for post indicator valves (PIVs) are in place, the OPEN/ SHUT signs are readable, and that cover glass is clean and in place.

- PIV bolts are tight and the barrel casings are intact.

- Main drain valves, auxiliary drains, and inspector's test valves are closed.

- FDC connection threads are unobstructed and in good condition, and caps are in place.

Dry-Pipe Systems

Dry-pipe systems have many elements in common with wet-pipe systems and are inspected in the same manner. However, these systems also have unique features that require special attention. During an inspection of a dry-pipe sprinkler system, inspectors should ensure that the following items are in operational condition:

- Indicating valves are open and properly supervised in the open position.

- Air-pressure readings correspond to previously recorded readings **(Figure 6.47)**.

- Ball-drip valves move freely and allow trapped water to seep out of the FDC.

- FDC threads are unobstructed and in good condition, and caps are in place.

- The system's air pressure is at 5 to 20 psi (35 kPa to 210 kPa) (depending on the system type) above the trip point and no air leaks are indicated by a rapid or steady air loss.

- The system's air compressor is well maintained, operable, and of sufficient size.

Caution
System air loss is often detected by the constant operation of the compressor or ultrasonic listening equipment.

Figure 6.46 Valve wheels and cranks should be tested to ensure they are in good condition.

Deluge and Preaction Systems

Deluge and preaction systems follow the same guidelines as those performed on wet and dry systems. More frequent inspections may be needed during cold weather. The supervisory air pressure in the preaction system should be checked weekly. The detection system should be inspected based on the type of detection method employed and the associated testing standards.

Figure 6.47 Air pressure should be determined and compared to previous readings. *Courtesy of the Sand Springs (OK) Fire Department.*

Summary

Automatic sprinkler systems and other water-based fire-suppression systems are essential to life safety and property conservation because of their efficiency and reliability. Numerous lives and countless dollars have been saved since sprinkler systems were first introduced over 100 years ago.

It is imperative that fire service personnel possess a good working knowledge of sprinkler systems, their components, and their operating methods. Inspection of systems to ensure proper operation and maintenance is an important aspect in the functioning of the automatic sprinkler system.

Review Questions

1. What are the benefits of automatic sprinkler systems?
2. What are the components of an automatic sprinkler system?
3. What is the difference between control valves and operating valves?
4. What determines the selection of the operating temperature of a sprinkler?
5. What are the types of specialty sprinklers?
6. What is the difference between a wet-pipe sprinkler system and a dry-pipe sprinkler system?
7. What is the difference between a deluge system and a preaction system?
8. What is the primary purpose of a residential sprinkler system?
9. What are the types of commodity classifications?
10. What are items that should be checked when testing and inspecting wet-pipe systems?

Standpipe and Hose Systems

Chapter Contents

Divider page photo courtesy of Texas Fire Protection Specialists

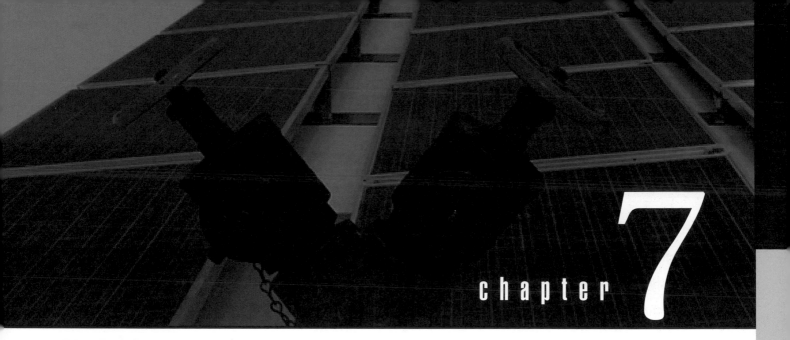

chapter **7**

Key Terms

FESHE Outcomes

Fire and Emergency Services Higher Education (FESHE) Outcomes: Fire Protection Systems

1. Explain the benefits of fire protection systems in various types of structures.

4. Identify the different types and components of sprinkler, standpipe, and foam systems.

10. Discuss the appropriate application of fire protection systems.

Standpipe and Hose Systems

Learning Objectives

After reading this chapter, students will be able to:

1. List the basic components of the standpipe system.

2. Identify the different classes of standpipe systems and their intended uses.

3. Describe the different types of standpipe systems.

4. Discuss the function of a fire department connection (FDC).

5. Discuss considerations concerning water supply and water pressure for standpipe and hose systems.

6. Describe basic categories of pressure-regulating devices.

7. Describe the process for initial and in-service inspections of standpipes.

8. Discuss fire department operations and the use of standpipe and hose systems.

Chapter 7
Standpipe and Hose Systems

Case History

In February of 1991, a fire was reported at a high rise structure in Pennsylvania. Upon arrival, fire personnel were told by security guards that there was a fire on the 22nd floor. Firefighters made their way upstairs but were slowed due to a total loss of electricity in the building. When they reached the fire, firefighters connected to the building's standpipe system and began to attack the fire. Unfortunately, the water pressure was so low that firefighters were unable to make a successful attack. Fire department pumpers hooked up to the fire department connections for the standpipes in order to increase pressure, but were unable to do so.

The standpipe system of the building was initially designed to be used with smoothbore nozzles and pressure reducing valves at the standpipe outlets were set to approximately 60 psi (420 kPa). However, fire department personnel were equipped with fog nozzles that required 100 psi (700 kPa) to generate adequate fire flow. The pressure reducing valves were able to be adjusted in the field to allow more pressure, but required a special tool that was not available. Eventually, a system technician arrived with the adjusting tool and was able to adjust the valves to allow for the correct pressure. The fire lasted for more than 14 hours until it was brought under control by an automatic sprinkler system on the 30th floor. Unfortunately, 3 firefighters died while trying to ventilate the structure.

Standpipe and hose systems are designed to provide a means for rapidly deploying fire hoses and operating fire streams at locations that are remote from the fire apparatus. Standpipes can be found in a variety of buildings and structures where required by the authority having jurisdiction (AHJ).

The value of standpipes in expansive one-story structures is primarily one of expediency. Horizontal standpipes expedite fire control by reducing the time and effort needed to manually advance a hoseline several hundred feet (meters) to reach the seat of the fire. Horizontal standpipes can also facilitate the overhaul of fires that have been controlled by sprinkler systems by reducing the amount of hose needed to reach the area. In many high-rise buildings, a standpipe is the primary means for manual extinguishment and overhaul of

Figure 7.1 Standpipes are widely used in high-rise structures to provide water for fire fighting purposes.

Standpipe System — Wet or dry system of pipes in a large single-story or multistory building with fire hose outlets installed in different areas or on different levels of a building to be used by firefighters and/ or building occupants. The system is used to provide for quick deployment of hoselines during fire fighting operations.

a fire and is an essential aspect of the building's design **(Figure 7.1)**. Because the fire pump discharge pressure necessary to reach the top floor of a 50-story building may be 350 to 400 psi (2 450 kPa to 2 800 kPa), the proper operation of the standpipe system is critical.

A standpipe system may be used by firefighters, properly trained occupants, or both, depending on the type of system installed. The system may be supplied by a reliable water supply and/or augmented by water from fire department engines through a fire department connection (FDC) **(Figure 7.2)**. The standpipe system may also be part of or separate from an automatic sprinkler, water spray, water mist, or foam-water system.

Although **standpipe systems** are required in many buildings, they do not take the place of automatic sprinkler systems, nor do they lessen the need for sprinklers. Automatic sprinklers continue to be the most effective method of fire control.

Figure 7.2 Fire department apparatus can provide water and increase water pressure in the standpipe system through the FDC.

Figure 7.3 Water-flow control valves are commonly found at standpipe outlets.

Components and Classifications of Standpipe Systems

Components found in standpipe systems commonly include the following:

- Hose stations (the type and diameter of hose will be governed by the classification of the system)
- Water supply
- Water-flow control valves **(Figure 7.3)**
- Risers (piping systems used to transfer water from the supply to the discharge)
- Pressure-regulating devices
- Fire department connection (FDC)

NFPA® 14, *Standard for the Installation of Standpipes and Hose Systems,* is the standard that is used for the design and installation of standpipes. This standard establishes three classes of standpipe systems. These classifications are based on the intended use of the hose stations or discharge outlets. NFPA®13, *Standard for the Installation of Sprinkler Systems*, also contains information on hose stations.

Class I

Class I standpipe systems are primarily for use by fire-suppression personnel trained in handling large hoselines. Class I systems must be capable of supplying effective fire streams during the more advanced stages of fire within a building. A Class I system provides 2½ inch (65 mm) hose connections or hose stations attached to the standpipe riser **(Figure 7.4)**. The 2½ inch (65 mm) hose connections may be equipped with a **reducer** on the cap that allows for the connection of a 1½ inch (38 mm) hose coupling as well.

Class II

The Class II system is primarily designed for use by building occupants who are trained in its use or by fire department personnel. These systems are equipped with 1½ inch (38 mm) hose and nozzle and stored on a hose rack system. The hose used in these systems is typically a single-jacket type and equipped with a lightweight, twist-type shut-off nozzle. These systems are sometimes referred to as **house lines (Figure 7.5)**.

There is some disagreement over the value of a Class II system. The presence of the small hose may give a false sense of security to building occupants and create the impression that they should attempt to fight a fire even though the safer course would be to escape. Fire department personnel should follow their department's standard operating procedures (SOPs) concerning the use of these systems.

Class III

A Class III system combines the features of Class I and Class II systems. Class III systems provide 1½ inch (38 mm) hose stations to supply water for use by building occupants who have been trained and 2½ inch (65 mm) hose connections to supply a larger volume of water for use by fire departments and those trained in handling heavy fire streams **(Figure 7.6)**. The design of the system must allow both Class I and Class II services to be used simultaneously. It is possible that the local jurisdiction may request the removal of the hose, nozzles, and rack leaving only the 2½ inch (65 mm) and 1½ inch (38 mm) discharge connections **(Figure 7.7)**.

Figure 7.4 A typical Class I standpipe.

Reducer — Adapter used to attach a smaller hose to a larger hose. The female end has the larger threads, while the male end has the smaller threads.

House Line — Permanently fixed, private standpipe hoseline.

Figure 7.5 Class II standpipes include a hose and nozzle and are primarily designed for use by building occupants.

Figure 7.6 Class III standpipes combine features of Class I and Class II systems.

Figure 7.7 Some jurisdictions have removed hose and nozzle components of Class III systems.

Types of Standpipe Systems

Within the three classes of standpipe systems, there are different types. The different types of standpipe systems include the following:

- *Automatic wet* — This system contains water at all times. The water supply is capable of meeting the system demand automatically. The water-supply control valve is open, and pressure is maintained in the system at all times **(Figure 7.8)**. A wet standpipe with an automatic water supply is most desirable because water is constantly available at the hose station. Wet standpipe systems cannot be used in cold environments.

- *Automatic dry* — This system contains air under pressure to supervise the integrity of the piping. Water is admitted to the system through a dry pipe valve upon the opening of a hose valve. Automatic dry systems have a permanently attached water supply. This type of system has the disadvantage of greater cost and maintenance requirements.

- *Semiautomatic dry* — This is a standpipe system that is attached to a water supply that is capable of supplying the system demand at all times. It requires activation of a control device to provide water at hose connections. The system is arranged to admit water into the system when a dry-pipe valve is activated at the hose station.

- *Manual dry* — This system does not have a permanent water supply. It is designed to have water only when the system is being utilized through the fire department connection (FDC).

- *Manual wet* — There is no permanent water supply for this system, and water must be provided by the fire department. It is maintained full of water from a small source for the purpose of detecting leaks in the system.

Figure 7.8 Automatic wet standpipe systems contain water under pressure at all times.

Fire Department Connection

Each Class I or Class III standpipe system requires one or more FDCs through which the fire department engine can supply water into the system. A FDC allows the fire department to pump supplemental water into a standpipe system or sprinkler system. The FDC is located in an accessible area on the exterior of the structure or away from the building **(Figures 7.9 a and b)**. High-rise buildings having two or more zones require a FDC for each zone.

Figure 7.9a This FDC is attached to the structure.

Figure 7.9b Some FDCs are located away from the structure in a location that is easily accessible to fire department apparatus.

In high-rise buildings with multiple zones, the upper zones may be beyond the height to which a fire engine can effectively supply water. This height is usually around 450 feet (135 m), depending on the available hydrant pressure and other factors. For standpipe system zones beyond that height, a FDC is of no value unless the fire department's apparatus is equipped with a special high-pressure pump and hose, and the system has high-pressure piping.

Standard requirements specify that there shall be no shutoff valve between the FDC and the standpipe riser. In multiple-riser systems, however, gate valves are provided at the base of the individual risers to allow for maintenance while leaving the remainder of the risers operational.

Each FDC is required to have at least two 2½ inch (65 mm) connections or more as prescribed by the AHJ for fire department use **(Figure 7.10)**. In addition, there should be at least one 2½ inch (65 mm) connection for each 250 gallons per minute (gpm) (1 000 L/min) of system demand. The hose connections to the FDC typically have a female connection with National Hose Standard (NHS) threads and should be equipped with standard cap plugs, approved breakaway covers, or locking caps **(Figure 7.11)**. If the AHJ does not use NHS-type threads, it is important that the hose-connection threads conform to those used by the local fire department. Some jurisdictions require Storz-type couplings that allow large diameter hose to supply standpipe systems.

A sign with the word *STANDPIPE* on a plate or fitting must indicate the FDC **(Figure 7.12)**. If the FDC serves both the sprinkler and standpipe systems (combination system), the sign should state *AUTOSPKR AND STANDPIPE*. If the FDC does not service the entire building, the sign must specify which floors are serviced **(Figure 7.13)**. In some instances, the sign should also indicate the pressure required to meet system demand.

Figure 7.10 Some FDCs incorporate more than two 2½ inch (65 mm) connections.

Figure 7.11 Cap plugs for FDCs are easy to remove in an emergency and prevent debris from entering the opening. *Courtesy of the McKinney (TX) Fire Department.*

Figure 7.12 FDCs that serve only the standpipe must say STANDPIPE on the identification plate.

Figure 7.13 The information plate must identify the specific areas that are served if the FDC does not serve the entire building.

Water Considerations

Two different factors to consider with regards to the use of water with stand-pipe systems are water supply and water pressure. These two factors must be taken into account when designing and installing standpipe and hose systems.

Water Supply

The water supply for a standpipe system may come from a variety of sources, including public water supplies, pressure tanks, and gravity tanks. Not all of these water sources are practical in every situation. Water supplies can be used in combination with automatic or manual fire pumps for additional pressure.

The amount of water required for a standpipe system depends on the size and number of fire streams that are needed and the probable length of time the standpipe will be used. Water-supply requirements for standpipes are influenced by the size and occupancy of the building as well as the fuel loads and hazards present.

The water supply for Class I and Class III standpipe systems should provide 500 gpm (2 000 L/min) for at least 30 minutes with a residual pressure of 100 psi (700 kPa) at the most hydraulically remote 2½ inch (65 mm) outlet. A minimum of 65 psi (450 kPa) is required for the most remote 1½ (38 mm) inch outlet.

If more than one standpipe riser is needed to protect a building, the water supply must provide 250 gpm (1 000 L/min) for each additional riser to a maximum of 1,250 gpm (5 000 L/min) for an unsprinklered building and 1,000 gpm (4 000 L/min) for a sprinklered building. For Class II standpipes, 100 gpm (375 L/min) must be provided for at least 30 minutes with a residual pressure of at least 65 psi (450 kPa) at the highest outlet.

The current NFPA® 14 minimum requirement for residual pressure is 65 psi (450 kPa) for 1½ inch (38 mm) hose connections and 100 psi (700 kPa) for 2½ inch (65 mm) hose connections. However, these are minimums and may not be adequate in all instances. For example more pressure may be needed to supply a fog nozzle on the end of a 100-foot (30 m) hose connected to the topmost hose outlet. Because of this situation, some building and fire codes require higher minimum residual pressures. The code used by the AHJ should be consulted to determine the minimum residual pressures for standpipe-protected occupancies.

High-Rise Buildings

The height of the building and the class of service determine the size of the standpipe riser. For Class I and Class III service, the minimum riser is 4 inches (100 mm) for building heights less than 100 feet (30 m) and 6 inches (150 mm) for heights over 100 feet (30 m). When a Class I or Class III standpipe exceeds 100 feet (30 m) in height, the top 100 feet (30 m) is allowed to be 4-inch (100 mm) pipe. Standpipes that are part of a combined system (those that include a sprinkler and standpipe) are 6 inches (150 mm). Standpipes can also be sized hydraulically to provide the minimum required pressure at the topmost outlet. For Class II service, a riser should be 2 inches (50 mm) for a building height less than 50 feet (15 m). For a building over 50 feet (15 m) in height, the minimum size riser is 2½ inches (65 mm).

Current system design practice is to locate standpipes so that any part of a floor is within 130 feet (40 m) of the standpipe hose connection **(Figure 7.14)**. This distance allows any fire to be reached with 100 feet (30 m) of hose, plus a 30-foot (10 m) fire stream. Standpipes and their connections are most commonly located within noncombustible fire-rated stair enclosures so that firefighters have a protected point from which to begin an attack. If the building is so large that the standpipes located in the stairwells cannot provide coverage to the entire floor, additional stations or risers must be provided. Some buildings with rated horizontal exits will have a standpipe on each side of the opening.

In buildings with combined standpipe and sprinkler systems, the minimum riser size is 6 inches (150 mm). However, this requirement may not be applicable if the building is protected by an automatic sprinkler system and the combined system is hydraulically calculated to ensure that all water-supply requirements can be met.

The actual standpipe hose connections can be located not less than 3 feet (0.9 m) and not more than 5 feet (1.5 m) from floor level. These connections should be plainly visible and should not be obstructed. Any caps over the connections should be easy to remove.

Buildings equipped with a Class I or Class III system may be required to have a 2½ inch (65 mm) outlet on the roof **(Figure 7.15)**. This outlet may be required when any of the following situations are present:

- The building has a combustible roof.
- The building has a combustible structure or equipment on the roof.
- The building has exposures that present a fire hazard.

Figure 7.14 Standpipes may need to be placed midfloor in a long corridor to allow any area of the floor to be reached with a 100 foot (30 m) section of hose.

Figure 7.15 Roof standpipe outlets may be necessary if combustible structures or equipment are present on the roof. *Courtesy of Texas Fire Protection Specialists.*

Water Pressure

One problem encountered with standpipes installed in high-rise buildings is the increase in water-pressure requirements. This is a result of the height of the building. It is important to remember that 0.433 psi (3 kPa) is required to raise water 1 foot (0.3 m). When a building is 10- or 20-stories tall, the pressure on the system at the lower floors is so great that it makes hoses difficult to handle when they are attached to the hose stations on those floors. In these cases, pressure-reducing valves may be required to keep operating pressures below 175 psi (1 225 kPa) **(Figure 7.16, p. 178)**.

High-rise buildings may have several zones **(Figure 7.17, p. 178)**. For example, The Willis Tower (formerly Sears Tower) in Chicago is 1,400 feet (425 m) tall and has a standpipe system divided into 7 zones. Water for the upper zones is supplied by an individual-zone fire pump that is supplied directly from a water main, or it may receive water from the fire pump that supplies the lower zone. When one pump takes in water from the discharge of another, the pumps are said to be arranged in *series*.

Figure 7.16
Pressure reducing valves are sometimes necessary when standpipes are used on lower floors in order to reduce pressure at the outlet to a more manageable level.

Figure 7.17 Multiple zones may exist in larger buildings.

In high-rise buildings with several zones, the upper-zone pumps may be arranged to draft from tanks on the upper floors. The tanks, which hold several thousand gallons (Liters) of water, are filled automatically from lower-zone fire pumps and/or the domestic water-supply pumps by means of automatic float valves. These float valves open when the level in the tank begins to drop. Tanks located on the upper floors for a source of water supply provide for great system reliability. Should the lower-zone pumps fail, there is still water available to the pumps located on the upper floors of the building.

NFPA® standards require a pressure-regulating device at a hose outlet that exceeds 100 psi (700 kPa) for a 1½ inch (38 mm) connection and 175 psi (1 225 kPa) for a 2½ inch (65 mm) connection. This device will limit the pressure to 100 psi (700 kPa) unless the fire department has approved otherwise. Pressure-regulating devices prevent pressures that make hoses difficult or dangerous to handle. They also enhance system reliability because individual zones are extended to greater heights. Although this may make system design more complex, system economy can be improved by eliminating the number of pumps needed.

Three basic categories of pressure-regulating devices include the following:

- *Pressure-restricting devices* — Consist of a simple restricting orifice inserted into the waterway. The amount of pressure drop through the orifice plate depends on the orifice diameter and available flow and pressure within the system. Each standpipe discharge connection is fitted with a restricting orifice with different sizes being required for each floor and application. Pressure-restricting devices are limited to systems with 1½ inch (38 mm) hose discharges and 175 psi (1 225 kPa) maximum pressure. These devices are not a preferred type because they do not control or reduce the water pressure in the system.

- *Pressure-control valves* — Preferred for managing excessive pressure and considered to be the most reliable method of pressure control. This is because they use a pitot tube and gauge to read the pressure and automatically reduce the flow through the discharge. Some of these devices are field adjustable, while others are preset at the factory.

- *Pressure-reducing valves* — Preferred for managing excessive pressure and uses a spring mechanism that compensates for variations in pressure. These mechanisms balance the available pressure within the system with the pressure required for hoseline use. Pressure-reducing valves are generally field-adjustable when a situation arises where too little pressure is being supplied.

As in the case history at the beginning of this chapter, if a pressure-regulating device is not properly installed or is not properly adjusted for the required inlet pressure, outlet pressure, and flow, the available flow may be greatly reduced and fire-fighting capabilities seriously impaired. It is important that manufacturer instructions for installation and adjustment be followed carefully.

Standpipe Inspection and Testing

In order to ensure both compliance with local codes and standpipe operability, standpipes should be inspected when they are first installed and periodically thereafter. The following sections highlight inspection and testing procedures for standpipes. In addition to NFPA® standards, local codes and ordinances should be consulted regarding standpipe system requirements and installations.

Initial Installation Inspection and Tests

A standpipe system is a significant component in a building's design. Before system installation, detailed design plans should be submitted to the AHJ for approval. It is during this phase where the plans are checked for compliance with the jurisdiction's current codes. As construction proceeds, the installation of the standpipe will need to be evaluated. In high-rise buildings, the standpipe should be in partial operation as construction proceeds to provide protection during construction.

When installation is complete, the following tests and inspections should be performed:

- The system should be hydrostatically tested at a pressure of at least 200 psi (1 400 kPa) for 2 hours to ensure tightness and integrity of fittings. If the normal operating pressure is greater than 150 psi (1 050 kPa), the system should be tested at 50 psi (350 kPa) greater than its designed pressure.

- The system should be flow-tested to remove any construction debris and to ensure that there are no obstructions.

- On systems equipped with an automatic fire pump, a flow test should be performed at the highest outlet to ensure that the fire pump will start when the hose valve is opened.

- The system should be tested to ensure that it will deliver its rated flow and pressure.

- All devices should be inspected to ensure that they are listed by a nationally recognized testing laboratory such as Underwriters Laboratory Inc. (UL) or FM Global.

- Hose stations and connections should be checked to ensure that they are located in cases not less than 3 feet (0.9 m) and not greater than 5 feet (1.5 m) from the floor and are positioned so that the hose can be attached to the valve without kinking.

Figure 7.18 Hose threads should be inspected to ensure they are not damaged and that they match threads used by the fire department.

Figure 7.19 Locking FDC caps should be tested to ensure they can open easily. *Courtesy of the McKinney (TX) Fire Department.*

- Each hose cabinet or closet should be inspected for a conspicuous sign that reads *FIRE HOSE* and/or *FIRE HOSE FOR USE BY OCCUPANTS OF BUILDING.*

- FDCs should be checked for the proper fire department thread and proper signage.

- When a manual standpipe is installed, check for a sign indicating *MANUAL STANDPIPE FOR FIRE DEPARTMENT USE ONLY.*

CAUTION

Extra care should be taken during testing with flow drainage and discharge. This water is often extremely dirty and can easily stain exterior building finishes.

In-Service Inspections

As with all fire protection systems, standpipe systems need to be inspected and tested at regular intervals. A visual inspection should be made at least monthly by the building management. Because interior fire fighting depends on the standpipe system, the fire department should also inspect standpipes at regular intervals. Actual testing of standpipes should be conducted by a fire protection contractor or the building operating staff (if they are sufficiently knowledgeable). To avoid potential liability, it is prudent that fire department personnel not perform actual tests. However, fire department personnel should witness the testing procedures.

Fire department personnel should regularly inspect standpipe systems to ensure the following conditions are met:

- All water-supply valves are sealed in the open position.

- Power is available to the fire pump.

- Individual hose valves are free of paint, corrosion, and other impediments.

- Hose valve threads are not damaged and match fire department couplings **(Figure 7.18)**

- FDC and caps are in place and any locking caps can be opened **(Figure 7.19)**.

- Pipes are free of trash or debris.

- Hose valve wheels are present and not damaged.

- Hose cabinets are accessible.

- Hose is in good condition, has proper dryness, and is properly positioned on the rack.

- Hose nozzles are present and in good working order.

- Discharge outlets in a dry system are closed.

- Dry standpipes are drained of moisture.

- Access to the FDC and closest hydrant is not blocked.

- FDC is free of obstruction and the swivels rotate freely.

- Water-supply tanks are at the proper level.
- Any pressure-regulating devices are tested as required by the manufacturer.
- Hose valves on dry systems are closed.
- Dry systems are hydrostatically tested every 5 years.
- Regular testing of pressure-regulating devices is performed if installed.

Fire Department Operations

For standpipe systems to be effective, fire department personnel must be trained in their operation and understand the systems and their limitations. Most fire departments perform preplanning activities. As a part of this process, firefighters should locate the FDC, water-supply valves, pumps, tanks, and hose connections. Fire departments should also develop SOPs for the use of standpipe and hose systems.

Emergency responders must have the proper equipment to facilitate the use of the standpipe systems. Most departments have *standpipe packs* or *high-rise packs* on their apparatus **(Figure 7.20)**. Regardless of their name, these packs contain the equipment necessary to connect to the standpipe system. The equipment carried in these equipment packs may vary from jurisdiction to jurisdiction, but should include 100 to 200 feet (30 m to 60 m) of hose, a nozzle, and miscellaneous tools and wrenches.

The first fire-fighting crew on the scene of a building with a standpipe system should take this pack into the building when investigating the alarm or call. The next due engine should stand by or connect to the FDC to supply the system. When charging the system, the jurisdiction's SOPs must be followed. Firefighters should use a hose connection that is located in a protected stairway enclosure if possible as opposed to one that is located in a corridor or open area.

Figure 7.20 High-rise packs include the equipment necessary to deploy a hoseline from a standpipe and are readily available on the fire apparatus.

Summary

Standpipes and hose systems reduce the time and effort needed by fire-fighting personnel to attach hoselines and apply water during a fire emergency. In large or very tall buildings, much more effort and resources would be required to stretch hoselines if standpipes or FDCs were not already in place. Every minute counts during an emergency and the availability of these systems plays a crucial role in timely mitigation. Because it is easy to take built-in systems for granted, fire-fighting personnel must ensure that such systems are inspected and in good working order. There is no time to repair a system during an emergency.

Review Questions

1. What are the components of a standpipe system?

2. What are the three classes of standpipe systems?

3. What is the difference among the three classes of standpipe systems?

4. What are the five types of standpipe systems?

5. What is the purpose of a fire department connection (FDC)?

6. What factors influence the water-supply requirements for standpipes?

7. What are the current system design practices for locating standpipes in high-rise buildings?

8. Why do high-rise buildings require an increase in water pressure requirements?

9. What are the three basic categories of pressure-regulating devices?

10. What are items that fire department personnel should regularly inspect when inspecting standpipe systems?

Special Extinguishing Systems

Chapter Contents

Key Terms

FESHE Outcomes

Fire and Emergency Services Higher Education (FESHE) Outcomes: Fire Protection Systems

1. Explain the benefits of fire protection systems in various types of structures.

4. Identify the different types and components of sprinkler, standpipe, and foam systems.

6. Identify the different types of non-water based fire suppression systems.

10. Discuss the appropriate application of fire protection systems.

Special Extinguishing Systems

After reading this chapter, students will be able to:

1. Describe special extinguishing system classifications.

2. Discuss wet chemical fire-extinguishing systems.

3. Discuss dry chemical fire-extinguishing systems.

4. Discuss clean-agent fire-extinguishing systems.

5. Discuss carbon dioxide fire-extinguishing systems.

6. Describe water-mist fire-suppression systems.

7. Discuss the use of foam fire-extinguishing systems, types of foam, and foam generation.

8. Distinguish among types of foam systems.

9. Distinguish among characteristics of foam nozzles and sprinklers.

Chapter 8
Special Extinguishing Systems

Case History

In March of 2008, fire personnel were dispatched to a restaurant kitchen fire in Oklahoma. Upon arrival, firefighters found light smoke in the dining area and made their way to the kitchen. The fire in the kitchen had already been extinguished by the wet chemical fire-extinguishing system in the exhaust hood. Splattered grease had ignited accumulated debris in the exhaust hood, causing a fire. Restaurant personnel manually activated the extinguishing system and then evacuated the restaurant. Because of the quick actions of restaurant staff and the effectiveness of the wet chemical system, there were no injuries and the restaurant was able to reopen the next evening.

Special extinguishing systems are used in locations where water-based automatic sprinklers may not be the best or most reasonable solution to a fire problem. These locations might contain food-preparation equipment, combustible metals, or highly sensitive computer or electronic equipment. In these unique locations, water might contribute to the fire, cause additional damage to the building or its contents, or may not effectively stop the fire.

Unlike water-based systems, which in most applications have an almost unlimited supply of water, special extinguishing systems have a limited amount of extinguishing agent. Automatic sprinkler systems are typically designed to control the fire, whereas special-agent fire-extinguishing systems are normally intended to extinguish the fire. These systems are also typically designed for a specific hazard or location and a limited area of coverage.

This chapter discusses the various classifications and types of special extinguishing systems, including wet chemical, dry chemical, clean agents, carbon dioxide, water mist, and foam. In addition, it will discuss the operations, components, agents, and inspection and testing requirements of these systems.

Special Extinguishing System Classifications and Types

Extinguishing systems are classified based upon the class of fire or fires they can effectively extinguish. Fires have been broadly grouped into five classifications according to the burning characteristics of various combustible materials. These classifications include the following:

- *Class A fire* — Involves ordinary combustibles such as wood, cloth, paper, rubber, and many plastics **(Figure 8.1)**. These fires can be extinguished by cooling, smothering, insulating, or inhibiting the chemical chain reaction.

- *Class B fire* — Involves flammable or combustible liquids and gases, including greases and similar fuels, which can be extinguished by oxygen exclusion, smothering, insulating, and inhibiting the chemical chain reaction **(Figure 8.2)**.

- *Class C fire* — Involves energized electrical equipment, which requires the use of a nonconductive agent for protection of the operator. If the electrical power is eliminated, these fires become Class A or Class B and may be extinguished appropriately.

- *Class D fire* — Involves combustible metals such as magnesium, potassium, sodium, titanium, and zirconium, which require the use of an agent that absorbs heat and does not react with the burning metal.

- *Class K fire* — Involves cooking oils and fats in appreciable depth. Class K-rated agents work by forming a barrier over the product, thus smothering and cooling the fire.

Figure 8.1 Class A fires involve ordinary combustibles such as these wooden pallets.

Figure 8.2 Class B fires involve flammable and combustible liquids and gases.

Labels are affixed to the extinguishing system storage tank to indicate the class of fire for which the system is approved.

As mentioned earlier, special extinguishing systems are generally designed with a specific hazard or application in mind. In addition, these systems typically do not serve an entire occupancy, but instead are designed to protect a specific area. Common types of special extinguishing systems include the following:

- Wet Chemical Extinguishing Systems
- Dry Chemical Extinguishing Systems

- Clean-Agent Extinguishing Systems
- Carbon Dioxide Extinguishing Systems
- Water Mist Extinguishing Systems

Wet Chemical Fire-Extinguishing Systems

A **wet chemical fire-extinguishing system** is best suited for application in commercial cooking hoods, plenums, ducts, and associated cooking appliances **(Figure 8.3)**. These systems are used in situations where rapid fire knockdown is required. The wet-chemical system is most effective when used on fires in deep fat fryers. The nature of the chemical is such that it reacts with animal or vegetable oil and forms a soapy foam through a process called **saponification**. This foam forms a blanket that helps to prevent reignition by separating the fuel from oxygen. The wet chemical system should provide a foam blanket that remains intact at least 20 minutes and not splatter the grease or oil when discharged. When using a wet-chemical agent, grease or oil fires are extinguished by fuel removal, cooling, smothering, and flame inhibition.

Wet Chemical Fire-Extinguishing System — Extinguishing system that uses a wet chemical solution as the primary extinguishing agent; usually installed in range hoods and associated ducting where grease may accumulate.

Saponification — A phenomenon that occurs when mixtures of alkaline-based chemicals and certain cooking oils come into contact, resulting in the formation of a soapy film.

Wet Chemical Agents

Wet chemical extinguishing agents are typically composed of water and either potassium carbonate, potassium citrate, or potassium acetate. The agent is delivered to the hazard area in the form of a spray. These are effective extinguishing agents for fires involving flammable/combustible liquids such as grease or oil or ordinary combustibles such as paper and wood.

More recently introduced alkaline mixtures are useful for attacking Class K fires because of their ability to generate a soapy, foam-like film. The agents used in this system are the same as those described for portable fire extinguishers in Chapter 9, Portable Fire Extinguishers.

Wet chemical systems can be messy, particularly when food greases are involved. The spray from a wet chemical system can also migrate to surrounding surfaces, causing corrosion of electrical wires or damage to hot

Figure 8.3 Wet chemical systems are typically found in commercial kitchen hoods. *Courtesy of the Sand Springs (OK) Fire Department.*

cooking equipment. For this reason, it is very important to ensure a prompt cleanup after activation of a wet chemical system. This type of system is not recommended for electrical fires because the spray mist may act as a conductor.

System Components

Wet chemical systems typically have the same components. These components are as follows:

- Storage tank or tanks for expellant gas and agent **(Figure 8.4, p. 190)**
- Piping to carry the gas and agent
- Nozzles to disperse the agent **(Figure 8.5, p. 190)**
- Actuating mechanism

Figure 8.4 Storage tanks for both the expellant gas and the extinguishing agent are located adjacent to the system.

Figure 8.5 Nozzles in a typical wet chemical system.

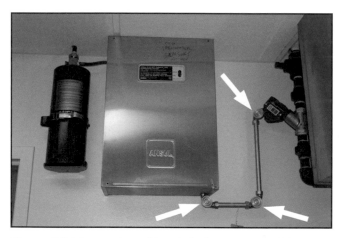

Figure 8.6 The number of bends in the system's piping must be taken into account when calculating piping requirements.

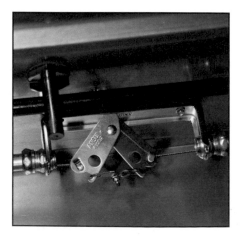

The storage container may contain both the agent and the pressurized expellant gas, or the agent and the gas may be stored separately. A pressure gauge attached to the container is an indication of a stored-pressure container. The expellant gas used is typically either nitrogen or carbon dioxide. System tanks must be located as close to the discharge point as possible, but they also must be in a temperature-controlled area.

The wet chemical agent is delivered to the hazard through nozzles attached to a system of fixed piping. The piping is specially designed to account for the unique flow characteristics of the agent. The proper pipe size, number of bends and fittings, and pressure drop (friction loss) are all taken into account when calculating piping requirements **(Figure 8.6)**.

Wet chemical is released into the piping system in response to activation devices. These devices are usually designed to activate when fusible links melt due to heat. The fusible links trigger a mechanical or electrical release that in turn starts the flow of expellant gas and agent **(Figure 8.7)**. Another method of automatic activation includes pressurized pneumatic tubing. Fixed systems should be capable of manual activation and must be equipped with automatic fuel or power shutoffs. The shutoff device must be restored manually. Detailed maintenance procedures and restoration of these systems should be left to trained personnel.

Inspection And Testing Procedures

A record should be kept of each inspection and any problems noted. Problems should be corrected immediately by the system service provider. Model codes require most systems to be inspected on a semiannual basis by a competent and trained individual. The system should be checked

Figure 8.7 Fusible links in the system are activated by heat and trigger a release that starts the flow of extinguishing agent.

to ensure that all components are working properly, piping is not obstructed, and there is no evidence of corrosion, structural damage, or repairs. The liquid levels in nonpressurized containers should also be checked. Trained personnel should be able to inspect these systems for the following:

- Mechanical damage
- Aim of nozzles
- Changes in hazards
- Proper pressures in stored-pressure containers
- Maintenance tags for scheduled service

Dry Chemical Fire-Extinguishing Systems

Dry chemical fire-extinguishing systems are used wherever rapid fire extinguishment is required and where reignition of the burning material is unlikely. These systems are most commonly used to protect the following areas:

- Flammable and combustible liquid storage rooms
- Dip tanks
- Paint spray booths **(Figure 8.8)**
- Older commercial cooking areas or kitchens
- Exhaust duct systems
- Heavy equipment, such as earthmovers and generators

All dry chemical systems should meet the requirements set forth in NFPA® 17, *Standard for Dry Chemical Extinguishing Systems*. The components of dry chemical systems are virtually the same as those for wet chemical systems. Two methods used for the application of dry chemical extinguishing agents are fixed systems and handheld hoselines, however only fixed systems will be addressed in this manual.

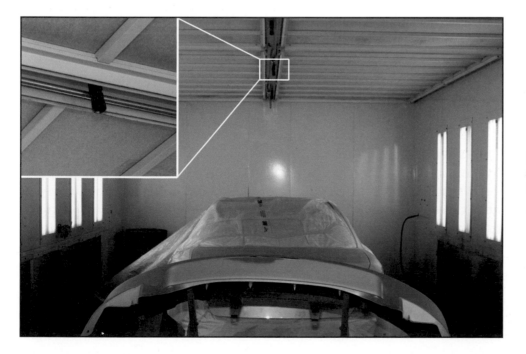

Figure 8.8 Dry chemical systems are commonly found in paint spray booths.

Fixed Dry Chemical Systems

Fixed dry chemical systems consist of the agent storage tanks, expellant storage tanks, a heat-detection and activation system, piping, and nozzles. Two main types of fixed systems are as follows:

1. *Local application* — This is the most common type of fixed system. Local application systems discharge agent onto a specific surface such as the cooking area in a restaurant kitchen. These systems are no longer code-compliant or listed for this use but may still be encountered.

2. *Total flooding* — This type of system introduces a thick concentration of agent into a closed area, such as a spray paint booth.

Dry Chemical Agents

Dry chemicals discharge a cloud of chemical that leaves a residue. This residue creates cleanup problems after system operation. A dry chemical system is not recommended for use in areas that contain sensitive electronic equipment. The chemical residue has insulating characteristics that hinder the operation of the equipment unless extensive cleanup is performed. The agent also becomes corrosive when exposed to moisture. Two dry chemical extinguishing agents that are currently used in dry chemical extinguishing systems include the following:

1. *Sodium bicarbonate* — Also known as *ordinary dry chemical*; sodium bicarbonate is effective on Class B and Class C fires. Sodium bicarbonate is twice as effective as an equal amount of carbon dioxide for Class B fires. This agent is also used for Class A fires in textile machinery where the fibers can produce a surface fire. Sodium bicarbonate used in fire-extinguishing systems is chemically treated to be water repellant and free-flowing.

2. *Monoammonium phosphate* — Also known as *multipurpose dry chemical*; monoammonium phosphate is effective on Class A, Class B, and Class C fires. This agent has an action similar to other dry chemicals on flammable liquid fires. Using a combination of extinguishing methods, it quickly extinguishes flaming combustion. When used on Class A materials, monoammonium phosphate melts, forming a solid coating that extinguishes the fire by smothering. Monoammonium phosphate is the most corrosive of dry chemical agents and can have a corrosive effect on unprotected metals. Corrosion can form around extinguisher system nozzles, piping, and agent containers.

Inspection and Testing Procedures

A record should be kept indicating that each inspection was made. Any problems that are noted during inspection should be corrected immediately. In most cases, correction of deficiencies requires notification of the fire protection system company that is responsible for the maintenance of the system. Inspections should be conducted as required by NFPA® 17. This standard has recommendations for monthly and semiannual inspection and testing.

Clean-Agent Fire-Extinguishing System — System that uses special extinguishing agents that leave little or no residue.

Clean-Agent Fire-Extinguishing Systems

Clean-agent fire-extinguishing systems are used in areas where wet or dry systems may be undesirable or unsuitable **(Figure 8.9)**. These areas must have undergone a room-integrity test prior to their use to ensure they are effective. Clean-agent systems store the extinguishing agent as a liquid. When the agent is exposed to the atmosphere, it turns to gas. In some cases, the gas displaces oxygen. In others, it disrupts the chemical chain reaction of the fire. Depending on the agent, the area may become untenable for occupants due to potentially lowered oxygen levels **(Figure 8.10)**. Clean agents are effective on Class A, Class B, and Class C fires and will not conduct electricity.

Some typical applications for clean-agent fire-extinguishing systems include the following:

- Computer rooms
- Telecommunications facilities
- Clean (manufacturing) rooms
- Data storage areas
- Irreplaceable document and art storage rooms
- Laboratories
- Art galleries
- Boats and vehicles

For flooding applications to be effective, room integrity is critical. In those areas where these systems are used, there are automatic door closers, door sweeps, and **predischarge warning devices (Figure 8.11)**. In addition, any heating, ventilating, and air-conditioning (HVAC) systems must be controlled. This allows the occupants in those areas to evacuate the space or room before the system discharges. Placarding or identification signage is extremely important in areas where these systems are installed.

Figure 8.9 This small clean-agent system is incorporated to protect a small records room.

Predischarge Warning Device — Alarm that sounds before a total flooding fire extinguishing system is about to discharge. This gives occupants the opportunity to leave the area.

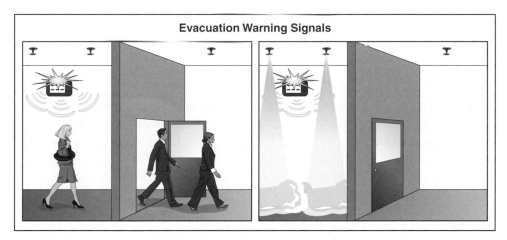

Figure 8.10 Because clean-agents can pose a hazard for occupants, an audible warning system is often incorporated.

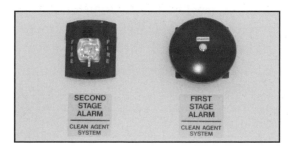

Figure 8.11 Predischarge warning devices for clean-agent release may include multiple stages. *Courtesy of Texas Fire Prevention Specialists*.

Clean Agents

Clean agents are in a general category of fire-extinguishing agents that leave no residue. These agents extinguish by smothering the burning material, excluding oxygen, and interrupting the chemical chain reaction of the fire. Clean agents are approved by the U.S. Environmental Protection Agency (EPA) as nonharmful to the atmosphere.

One of the first groups of clean agents developed were halogenated fire-extinguishing agents. Halogenated agents are principally effective on Class B and Class C fires. The word **halon** has been commonly used to describe this group of agents.

While halon agents are effective, they have been proven to be harmful to humans and the earth's ozone layer, so restrictions have been placed on their production. Although the Montreal Protocol of 1987 provided for a phase out of halon agents and forbade manufacture of new halon agents after January 1, 1994, limited production continues because of some exceptions to the phase out plan.

Two types of halons are still in use: Halon 1211 (bromochlorodifluoromethane) and Halon 1301 (bromotrifluormethane). Halon 1211 is most commonly found in portable extinguishers. Halon 1301 is used in some portable fire extinguishers but is more commonly found in fixed fire-extinguishing systems for total flooding applications.

Today there are products that extinguish fires in the same manner as halogenated extinguishing agents but do not cause damage to the atmosphere. These products are known as halon-replacement agents. Common halon-replacement agents include the following:

Figure 8.12 Inergen® is a commonly used halon-replacement agent.

- *Halotron®* — Clean agent hydrochlorofluorocarbon that, when discharged, is a rapidly evaporating liquid. Halotron® leaves no residue and meets Environmental Protection Agency (EPA) minimum standards for discharge into the atmosphere. The agent does not conduct electricity back to the operator, making it suitable for Class C fires. Halotron® has a limited Class A rating.

- *FM-200®* — Hydrofluorocarbon that is considered to be an acceptable alternative to Halon 1301 because it leaves no residue and is not harmful to humans and the environment. This agent does require significantly more agent for effective extinguishment than did Halon 1301. There are several derivatives of FM-200® that have slightly different compositions.

- *Inergen®* — Blend of three naturally occurring gases: nitrogen, argon, and carbon dioxide. It is stored in cylinders near the facility under protection (**Figure 8.12**). Inergen® is environmentally safe and does not contain a chemical composition like many other proposed halon alternatives.

- *Novec™ 1230* — Halon replacement that is stored in a liquid state but converts to a gas when discharged through a nozzle. Novec™ 1230 is environmentally sustainable and suited for sensitive areas such as computer rooms and museums.

- *Stat-X®* — Agent that is aerosolized and extinguishes a fire by interrupting the chemical chain reaction. Stat-X® is contained in canisters that release the product when activated. The agent is commonly used to protect sensitive electrical equipment and is also used in military combat vehicles.

Clean-Agent System Components

Clean-agent extinguishing systems may be fixed systems that are designed for local application or total flooding agent distribution. Components for these systems are similar to other systems previously discussed in this chapter. Clean-agent system components include actuation devices, agent storage containers, piping, and discharge nozzles.

Inspection and Testing Procedures

Inspection and testing procedures for clean-agent fire-extinguishing systems can be found in NFPA® 2001, *Standard on Clean Agent Fire Extinguishing Systems*. An annual inspection by qualified personnel is required for all system components. Semiannually, the quantity and pressure of the clean agent must be checked. Records on all tests and inspections must be maintained and made available for review by authority having jurisdiction (AHJ) inspectors.

In addition to system inspections and tests, the protected enclosure must be inspected annually. The inspector must determine that the integrity of the space has not been compromised by penetrations that would permit the agent to escape during a discharge.

Carbon Dioxide Fire-Extinguishing Systems

Carbon dioxide (CO_2) is a type of clean agent that has been proven effective for extinguishing most combustible materials fires with the exception of some active metals, metal hydrides, and other materials that contain available oxygen. The limitations of CO_2 are related to health effects associated with its use as well as restrictions imposed by the combustible material itself.

As delivered, CO_2 is extremely cold [approximately -110°F (-79°C)] and can freeze exposed skin. The agent, however, has a limited cooling effect on a fire. The primary mechanism of extinguishment of CO_2 is accomplished through oxygen removal or smothering. The cooling effects of its application (although small) are realized when the agent is applied directly to the burning material.

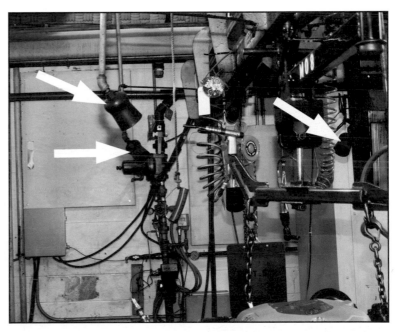

Figure 8.13 Carbon dioxide extinguishing systems are often designed to protect electrical equipment.

CO_2 fire extinguishing systems closely resemble clean-agent systems. All systems must adhere to the requirements as described in NFPA® 12, *Standard on Carbon Dioxide Extinguishing Systems*. These systems have been used to extinguish fires involving the following materials or equipment:

- Flammable and combustible liquids
- Electrical equipment and energized equipment **(Figure 8.13)**
- Flammable gases
- Other combustibles including cellulose materials

Although CO_2 systems were almost phased out by the growth of halon and subsequent clean-agent systems in the 1970s and 1980s, there has been a resurgence of these systems as the need to replace halon-based systems has increased. A CO_2 fire-extinguishing system can be used to protect a wide variety of hazards through total flooding or local fire-protection applications. Handheld hoseline and standpipe systems are also used, although they are not common. Some types of hazards that CO_2 can protect include the following:

- Automobile manufacturing facilities
- Industrial plants
- Refineries/chemical plants
- Paint and coating operations
- Food/agricultural processing plants
- Pharmaceutical manufacturing facilities
- Printing presses

The most serious problem involving CO_2 systems, especially total flooding systems, is personnel safety. The elimination of oxygen from a fire also eliminates breathable oxygen from the atmosphere, causing an asphyxiation hazard for personnel in the area. Total flooding systems must be provided with predischarge alarms as well as discharge alarms. A predischarge alarm notifies those present that the system is about to activate.

Figure 8.14 Carbon dioxide systems are stored at either high or low pressure. System pressure can be determined by reading the pressure gauges on the system.

Carbon Dioxide System Components

The components of the CO_2 system are similar to other special extinguishing systems. These components include actuation devices, agent storage containers, piping, and nozzles. The following are the three means of actuation for CO_2 systems:

- *Automatic operation* — Triggered by a product-of-combustion detector such as smoke detectors, fixed-temperature, or rate-of-rise detection equipment.

- *Normal manual operation* — Triggered by a person manually operating a control device and putting the system through its complete cycle of operation, including predischarge alarms.

- *Emergency manual operation* — Used only when the other two actuation modes fail, causing the system to discharge immediately and without any advance warning to individuals in the area.

CO_2 systems exist as either high-pressure systems or low-pressure systems. In a high-pressure system, the CO_2 is stored in standard U.S. Department of Transportation (DOT)-approved cylinders at a pressure of about 850 psi (5 860 kPa) **(Figure 8.14)**. A low-pressure system is designed to protect much larger hazards. The liquefied CO_2 in these systems is stored in large, refrigerated tanks at 300 psi (2 068 kPa) at a temperature of 0°F (-18°C).

In either a high-pressure or a low-pressure system, the containers are connected to the discharge nozzles through a system of fixed piping. Nozzles for total flooding systems may be the high- or low-velocity types. However, high-velocity nozzles promote better disbursement of the agent throughout the entire area. Local application nozzles are typically the low-velocity type, which reduces the possibility of splashing the burning product when it comes in contact with the agent.

Inspection and Testing

Due to the complexity of CO_2 systems, inspection items to be considered by fire inspectors or personnel should be limited to the following:

- Physical damage of the components
- Excessive corrosion
- Change in hazard
- Enclosure integrity
- Up-to-date test, maintenance, and inspection records

Water-Mist Fire-Suppression Systems

One system gaining more popularity in the field of fire protection is the water-mist system. While not considered an automatic sprinkler system, a water-mist fire-suppression system is very similar. This system sprays water onto a fire in a fine mist that absorbs larger quantities of heat than an automatic sprinkler system **(Figure 8.15)**. The fine mist of water controls or extinguishes the fire by displacing oxygen and blocking radiant heat production. A benefit of water-mist fire-suppression systems is the limited amount of water used compared to traditional sprinkler systems. These systems prevent and mitigate excessive water damage in occupancies where they are used.

Figure 8.15 Water mist nozzles are designed to release a fine spray of water. *Courtesy of Securiplex, LLC.*

The development of water-mist systems is relatively recent, and has been driven by the need for a replacement for halon and other oxygen-depleting systems. In addition, there has been a need to provide sprinkler protection on passenger cruise ships. NFPA® 750, *Standard on Water Mist Fire Protection Systems,* provides the requirements for installation of water-mist fire-suppression systems. Water-mist systems may be found in locations such as tunnels, passenger ships, aircraft passenger compartments, machinery spaces, and turbine enclosures. These applications are limited to the size of the area protected and can be applied to local application, total compartmentation, or zoned application.

In general, a water-mist system is composed of small-diameter, pressure-rated copper or stainless-steel tubing. Small-diameter spray nozzles are spaced evenly on the tubing. Depending on the design of the system, the spray nozzles may be of the open or closed sprinkler variety. Systems are activated by a product-of-combustion detection system. Water-mist systems should also have a means of manual operation, should the fire be discovered by individuals before the activation of the system.

Water-mist systems are designed to be operated at considerably higher pressures than standard sprinkler systems. These systems are designed for fire suppression, control, or extinguishment purposes and may have a set amount of time of discharge. Water-mist systems are divided into low [175 psi (1 200 kPa) or less], intermediate [175 to 500 psi (1 200 to 3 500 kPa)], or high [500 psi (3 500 kPa) or greater] pressure systems. Compressed-air, nitrogen, or high-pressure water pumps create these higher pressures. The most common type of water-mist system works similarly to a traditional deluge sprinkler system. All of the spray nozzles in a particular room or zone are open and when the detection devices activate, the water is allowed to discharge from the spray nozzles.

Water-mist systems may present an eye hazard during operation due to the high operating pressure of the system. These high-pressure systems may also present a noise hazard during discharge. Additionally, the activation of water-mist systems can result in reduced visibility, which may increase the time necessary for egress of occupants from the protected area.

Figure 8.16 An activated foam system. *Courtesy of Rolf Jensen & Associates.*

Foam Fire-Extinguishing Systems

A foam fire-extinguishing system is used when water alone may not be an effective fire-extinguishing agent **(Figure 8.16)**. In general, foam works by forming a blanket on the burning fuel. The foam blanket excludes oxygen, stops the burning process, and cools adjoining hot surfaces. The type of system and foam used depends largely on the hazard being protected. These hazards and locations include but are not limited to:

- Flammable liquids
- Processing or storage facilities
- Tire storage
- Aircraft hangars
- Rolled-paper or fabric-storage facilities

Foam extinguishes a fire by one or more methods including the following:

- **Smothering** — Prevents air and flammable vapors from combining.
- **Separating** — Intervenes between the fuel and the fire.
- **Cooling** — Lowers the temperature of the fuel and adjacent surfaces.
- **Suppressing** — Prevents the release of flammable vapors.

Standards and requirements for foam system design, placement, and other technical information can be found in NFPA® 11, *Standard for Low-, Medium-, and High-Expansion Foam*; NFPA® 16, *Standard for the Installation of Foam-Water Sprinkler and Foam-Water Spray Systems*; NFPA® 30, *Flammable and Combustible Liquids Code*; and NFPA® 409, *Standard on Aircraft Hangars*. Design specifications and requirements of foam sprinkler systems can also be found in publications from manufacturers and nationally recognized testing laboratories such as Underwriters Laboratories Inc. (UL) and FM Global.

Types of Foam

There are a number of types of foam agents available. Some foam concentrates are designed for specific applications, while others are suitable for all types of flammable liquids. The most common types of foam include the following:

- Aqueous film forming foam (AFFF)
- Fluoroprotein (FP)
- Film-forming fluoroprotein (FFFP)
- Protein (P)
- Alcohol-resistant (AR)
- Medium- and high-expansion foam

Foam Generation

Most fire-extinguishing foam concentrates in use today are of the mechanical type, which means they must be **proportioned** with water and mixed with air (*aerated*) before they can be used. Before discussing the foam-generating process, it is important to understand the following terms:

- ***Foam concentrate*** — Raw foam liquid before the introduction of water and air. It is usually shipped in 5-gallon (19 L) buckets or 55-gallon (200 L) drums but may be shipped in totes that contain several hundred gallons (liters) of product **(Figure 8.17)**. Concentrate can also be stored in large fixed tanks that can hold 500 gallons (1 900 L).

- ***Foam proportioner*** — Device that introduces the correct amount of foam concentrate into the water stream to make the foam solution.

- ***Foam solution*** — Homogeneous mixture of foam concentrate and water before the introduction of air.

- ***Foam (also known as finished foam)*** — Completed product once air is introduced into the foam solution.

Proportioning — The mixing of water with an appropriate amount of foam concentrate to form a foam solution.

Figure 8.17 Foam concentrate is available in a variety of concentrations and quantities. *Courtesy of Doddy Photography.*

Four elements are necessary to produce high-quality fire fighting foam: foam concentrate, water, air, and mechanical agitation. All of these elements must be present and introduced in the correct ratios. Removing any element will result in either no foam or an unusable liquid.

There are two stages in the formation of foam: First, water is mixed with the foam liquid concentrate to form a foam solution. This is known as the proportioning stage of foam production. Second, the foam solution passes through the piping or hoseline to a foam nozzle or sprinkler, and the foam nozzle or sprinkler **aerates** the foam solution to form finished foam.

An **eductor** is a type of **foam proportioner** that injects foam concentrate directly into the water flowing through a hose or pipe **(Figure 8.18)**. The eductor is designed to inject the correct amount of foam concentrate so that the desired consistency of foam solution is achieved. There are many different types of eductors and some of these types will be discussed later in the chapter.

Proportioning equipment and foam nozzles or sprinklers are engineered to work together. Using a foam proportioner that is not hydraulically matched to the foam nozzle or sprinkler, even if the two are made by the same manufacturer, can result in unsatisfactory foam or no foam at all.

Figure 8.18 This inline eductor injects foam concentrate directly into the hoseline at the desired amount. *Courtesy of Tom Hughes.*

Types of Foam Systems

A foam system must have an adequate water supply, a supply of foam concentrate, a piping system, proportioning equipment, and foam makers (or discharging devices). Each of the following systems are discussed in further detail in the sections that follow:

- Fixed
- Semifixed Type A
- Semifixed Type B
- High-Expansion
- Foam/Water

Fixed- Foam Systems

A **fixed-foam fire-extinguishing system** is a complete installation that is piped from a central foam station **(Figure 8.19)**. These systems automatically discharge foam through fixed delivery outlets to the protected hazard. If a pump is required to increase pressure in the system, it is usually permanently installed. Fixed systems may be the total-flooding or local-application type. Most fixed systems are deluge types that have unlimited water supplies and may also have large foam supplies. They require actuation by some sort of product-of-combustion detection system. Fixed systems use low-, medium-, or high-expansion foam.

Semifixed Type A Systems

In a semifixed Type A system, the foam discharge piping is in place but is not attached to a permanent source of foam. The semifixed Type A system requires a separate mobile foam-solution source, which is usually a fire brigade or fire department pumper **(Figure 8.20)**. This type of system is found in settings that involve several similar hazards, such as petroleum refineries, and are used primarily on flammable liquid storage tanks.

Systems such as these may utilize subsurface injection systems, where the foam is injected at the base of a burning storage tank and allowed to surface and extinguish the fire. These are used where topside application may not be effective because of wind or heavy fire conditions. Because the foam solution is lighter than the product in the tank, it floats to the top when introduced at the bottom of the tank. This method is highly effective for extinguishing bulk tank fires.

Semifixed Type B Systems

A semifixed Type B system provides a foam-solution source that is piped throughout a facility, much like a water distribution system. The foam solution is delivered to foam hydrants for connection to hoselines and portable foam application devices **(Figure 8.21, p. 202)**. The difference between a semifixed Type B system and a fixed system is that a fixed system actually applies foam to the hazards while the **semifixed system** merely provides foam capability to an area. In these systems, once the foam is provided to a certain location, it must then be applied manually.

> **Fixed-Foam Fire-Extinguishing System** — Complete installation of piping, foam concentrate storage, water supply, pumps, and delivery systems used to protect a specific hazard such as a petroleum storage facility.

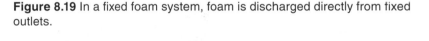
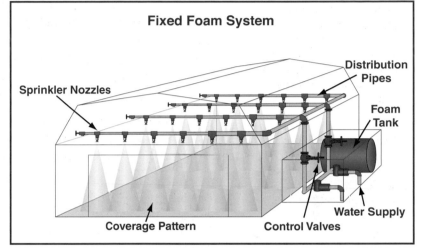

Fixed Foam System

Sprinkler Nozzles · Distribution Pipes · Foam Tank · Water Supply · Control Valves · Coverage Pattern

Figure 8.19 In a fixed foam system, foam is discharged directly from fixed outlets.

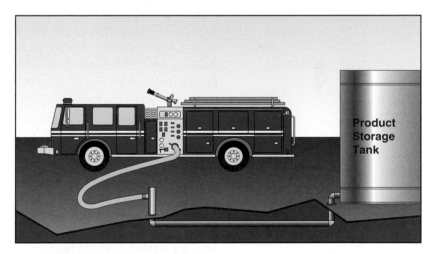

Figure 8.20 Semifixed Type A systems require foam to be supplied from a mobile fire apparatus.

Product Storage Tank

> **Semifixed System** — Foam system that is designed to provide fire-extinguishing capabilities to an area but is not automatic in operation and depends on human intervention to place it into operation.

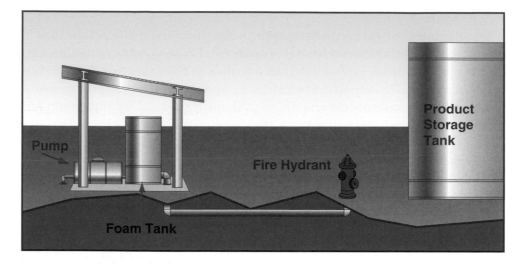

Figure 8.21 Semifixed Type B systems provide foam, typically through a hydrant that is distributed through hoselines and portable foam devices.

Product Storage Tank

Pump

Fire Hydrant

Foam Tank

Figure 8.22 High-expansion foam systems are used when total flooding applications are required, such as in aircraft hangars.

High-Expansion Foam Systems

A high-expansion foam system is designed for local application or total flooding in commercial and industrial applications **(Figure 8.22)**. These systems consist of the following primary components:

- Automatic detection or manual actuation systems
- Foam generator
- Piping from the water supply and foam concentrate storage tank to the generator

A high-expansion foam system can be actuated by any of the common fire detection devices or by a manual pull station. The foam generators are powered by electric or gasoline motors or by water. The generators should have a fresh-air intake to make sure that foam does not become contaminated by products of combustion. In addition, venting should be provided ahead of the foam to allow it to move through the area to be protected. In the total flooding application, the entire building can be filled to several feet (meters) above the highest storage area or equipment within a few minutes.

Foam/Water Systems

A foam/water system is similar to a deluge sprinkler system but has foam capabilities. These systems are used where there is a limited foam-concentrate supply but an unlimited supply of water. Thus, if the foam-concentrate supply becomes depleted, the system will continue to operate as a straight deluge sprinkler system.

Typically, the foam/water system is an automatic system that operates in the same manner as a regular deluge system. The major difference is that the foam-induction system and special aerating sprinklers are at the end of the piping. This type of system produces a *lean* foam solution that eventually expands six to eight times when it is discharged from the sprinkler. This produces a very fluid foam that will flow around obstructions after it is delivered.

The entire foam/water system may be divided into two parts: water system and foam system. The water system components are the same as those described in Chapter 6, Automatic Sprinkler Systems, for deluge sprinklers. The foam system contains a concentrate tank, a pump, a metering valve, a strainer, piping, and an actuation unit.

Protein, fluoroprotein (FP), and aqueous film forming foam (AFFF) concentrates may be used in foam/water systems. AFFF may also be discharged through regular water sprinklers with favorable results. When discharged through standard sprinklers, AFFF has greater velocity than when it is discharged through foam sprinklers. This tends to improve the spray and the penetration.

A foam/water system operates when the initiation devices sense the presence of fire and send an appropriate signal to the system control unit. This in turn triggers the deluge valve and the system. At the same time, the deluge valve on the foam side of the system opens, the foam pump starts, and the concentrate is introduced into the water flow. The foam/water solution flows to the sprinklers where air is introduced to create the foam. The foam is then delivered to the target area.

After the fire is extinguished, the system must be shut down, drained, and thoroughly flushed to remove any foam residue. The concentrate tanks must then be refilled. Once this is complete, the valves can be reset and the system restored to service.

Foam Nozzles and Sprinklers

Foam nozzles and sprinklers, sometimes referred to as *foam makers*, are the devices that deliver the foam to the fire or spill. Fixed systems may have both handlines and foam sprinklers or generators attached to them. Standard fixed-flow or automatic water fog nozzles may be used on AFFF or film-forming fluoroprotein (FFFP) handlines. The various types of foam nozzles and sprinklers are listed as follows:

- Smoothbore nozzle
- Low-expansion foam nozzle
- Self-educting foam nozzle
- Standard fixed-flow fog nozzle
- Automatic nozzle
- Foam/water sprinkler
- High back-pressure foam aspirator
- High-expansion foam generator

Smoothbore Nozzle

The use of smoothbore nozzles is limited to Class A, **compressed-air foam system (CAFS)** applications. In these applications, the smoothbore nozzle provides an effective fire stream that has maximum reach capabilities. Tests indicate that the reach of the CAFS fire stream can be more than twice the reach of a low-energy fire stream.

Compressed-Air Foam System (CAFS) — Generic term used to describe a high-energy foam-generation system consisting of an air compressor (or other air source), a water pump, and foam solution that injects air into the foam solution before it enters a hoseline.

When using a smoothbore nozzle with a CAFS, disregard the standard rule of thumb that the discharge orifice of the nozzle be no greater than one-half the diameter of the hose. Tests show that a 1½ inch (38 mm) hoseline may be equipped with a nozzle tip up to 1¼ inch (32 mm) in diameter and still provide an effective fire stream.

Low-Expansion Foam Nozzle

Foam solution must be mixed with air to form foam. The most effective appliance for generating low-expansion foam is the *air-aspirating* foam nozzle. The special design of a foam nozzle aerates the foam solution to provide the highest quality foam possible. Smaller foam nozzles may be handheld. Larger foam nozzles may be monitor-mounted units.

The fog/foam-type nozzle is marketed with two stream-shaping foam adapters. The basic adapter breaks the foam solution into small streams and at the same time inducts air through the **venturi effect (Figure 8.23)**. The cone-shaped attachment gives the nozzle extra reach, and the screen produces a more homogenous high-air-content foam for gentle applications **(Figure 8.24)**.

Figure 8.23 The basic adapter is attached to an adjustable fog nozzle and delivers the finished foam in small streams. *Courtesy of Tom Hughes.*

Figure 8.24 The cone adapter allows for greater reach with the foam stream and incorporates more air into the foam solution, allowing for gentle application. *Courtesy of Tom Hughes.*

Self-Educting Foam Nozzle

The **self-educting foam nozzle** operates with an eductor that is built into the nozzle rather than into the hoseline. As a result, its use requires the foam concentrate to be available where the nozzle is operated. The self-educting foam nozzle has the following advantages:

- It is easy to use.
- It is inexpensive.
- It works with lower pressures.
- A wide variety of flow rate versions are available.

Disadvantages include the logistical problems of relocation of the nozzle and the resulting need for relocation of the concentrate. Use of a foam nozzle also compromises firefighter safety in that personnel cannot move quickly, and they must leave the concentrate behind if they are required to back out of the area.

Standard Fixed-Flow Fog Nozzle

The fixed-flow, variable-pattern fog nozzle is used with foam solution to produce a low-expansion, short-lasting foam. This nozzle breaks the foam solution into tiny droplets and uses the agitation of water droplets moving through air to achieve its foaming action. The best application of this nozzle is when it is used with regular AFFF concentrate and Class A foams because its filming characteristic does not require a high-quality foam to be effective. These nozzles cannot be used with protein and FP foams or any alcohol-type foams. The fixed-flow fog nozzle may be used with alcohol-resistant AFFF foams on hydrocarbon fires but should not be used on polar solvent fires.

Automatic Nozzle

An automatic nozzle operates with an eductor in the same way that a fixed-flow fog nozzle operates. However, the eductor must be operated at the inlet pressure for which it was designed, and the nozzle must be fully open. An automatic nozzle may cause problems if the eductor is operated at a lower pressure than that recommended by the manufacturer or if the nozzle is not fully open.

Foam/Water Sprinkler

Foam/water sprinklers are found on fixed-foam deluge and foam/water systems. These systems use AFFF and FFFP foams. Many foam/water sprinklers resemble air-aspirating nozzles in that they use the venturi effect to mix air into the foam solution **(Figure 8.25)**. Some systems use standard sprinklers to form a less-expanded foam through simple turbulence of the water droplets falling through the air. Foam/water sprinklers come in upright and pendant designs. Their deflectors must be adapted to meet the specific installation requirements.

Figure 8.25 Foam/water sprinklers introduce air to the foam solution at the nozzle. *Courtesy of the United States Department of Defense.*

High Back Pressure Foam Aspirator

High back pressure foam aspirators, also known as *forcing foam makers*, are used to deliver foam under pressure. These foam makers are most commonly used in subsurface injection systems that protect cone-roof hydrocarbon storage tanks, but they are also used in other applications. High back pressure aspirators supply air direction to the foam solution through venturi action. This action typically produces a low-air-content foam that is homogenous and stable.

High-Expansion Foam Generator

High-expansion foam generators produce high air content, semistable foam. There are two basic types of high-expansion foam generators: water-aspirating type nozzle and mechanical blower. The water-aspirating type is very similar to other foam-producing nozzles except that it is much larger and longer. The back of the nozzle is open to allow airflow. The foam solution is pumped through the nozzle in a fine spray that mixes with air to form a moderate-expansion foam. The end of the nozzle has a screen or series of screens that breaks up the foam and further mixes it with air. These nozzles typically produce a lower air-volume foam than do mechanical blower generators.

A mechanical blower generator is similar to a smoke ejector in appearance **(Figure 8.26)**. It operates on the same principle as the water-aspirating type nozzle except that the air is forced through the foam spray instead of being pulled through by water movement. This device produces a higher air content foam and is typically associated with total flooding applications.

Figure 8.26 Mechanical blower generators closely resemble smoke ejectors because they incorporate a fan system.

Summary

Special fire-extinguishing systems are needed when the application of water is not the best way to extinguish a fire, either because of the nature of the hazard itself or because large amounts of water will ruin specialized equipment. In these situations, the specialized and controlled application of wet or dry chemicals, CO_2, clean agents, or foam are better choices. Because these systems are often custom-designed and all contain only a specific amount of extinguishing agent, they must be installed carefully and inspected regularly to ensure successful operation. Many of the chemicals and CO_2 used in specialized systems are dangerous to health, so it is imperative that people who may be affected be made aware of the potential hazards. Individuals must know how to exit an area promptly and activate the fire-extinguishing system from a position of safety.

Review Questions

1. What are common types of special extinguishing systems?

2. When would wet chemical fire-extinguishing systems be used?

3. When would dry chemical fire-extinguishing systems be used?

4. What are some typical applications for clean-agent fire-extinguishing systems?

5. What types of hazards do carbon dioxide fire extinguishing systems protect?

6. How does a water-mist fire-suppression system suppress or extinguish a fire?

7. What elements are necessary to produce high-quality fire fighting foam?

8. What are the types of foam systems?

9. What are the types of foam nozzles?

10. How does a high back pressure foam aspirator deliver foam?

Portable Fire Extinguishers

Chapter Contents

Key Terms

FESHE Outcomes

Fire and Emergency Services Higher Education (FESHE) Outcomes: Fire Protection Systems

11. Explain the operation and appropriate application for the different types of portable fire protection systems.

Portable Fire Extinguishers

After reading this chapter, students will be able to:

1. Identify classifications of portable fire extinguishers.

2. Describe the different symbol systems used on portable fire extinguishers.

3. Describe the method by which portable fire extinguishers are rated.

4. Identify the types of extinguishing agents used in portable fire extinguishers.

5. Distinguish among the advantages and disadvantages of different extinguishing agents.

6. Identify life safety hazards associated with certain extinguishing agents.

7. Describe the operating principles of different types of portable fire extinguishers.

8. Summarize considerations for the selection and distribution of portable fire extinguishers.

9. Summarize considerations for the installation and placement of portable fire extinguishers.

10. Discuss inspection, maintenance, and recharging procedures for portable fire extinguishers.

11. Describe hydrostatic testing of portable fire extinguishers.

12. Describe the general techniques for the use of portable fire extinguishers on different classes of fire.

Chapter 9
Portable Fire Extinguishers

Case History

Fire personnel were called to a Missouri big-box store in 2010 with the report of fire and smoke in the building. Upon arrival, firefighters encountered smoke but no fire. The fire, which involved an aisle of paper products, was extinguished by an employee using a fire extinguisher. Because the extinguisher was properly located and used, the fire was extinguished before it could spread to other areas of the store. Investigators ruled the cause of the fire to be arson.

Portable fire extinguishers are one of the most common fire protection devices found today. They can be found in fixed facilities, such as homes and businesses, and in vehicles, including passenger cars, boats, and fire apparatus. Portable fire extinguishers are intended for use on small fires in their incipient or early growth stages. In many cases, a portable extinguisher can control or extinguish a small fire in much less time than it would take for firefighters to deploy a hoseline. In some fire departments, one member on the initial attack team in a high-rise fire carries either a water type or a multipurpose portable extinguisher. If the fire proves to still be small upon reaching it, the extinguisher is used to extinguish the fire.

Many view fire extinguishers as the first line of defense against incipient fires. However, a fire extinguisher should not be viewed as a substitute for automatic fire detection and suppression systems. Rather it should be seen as a complement to these systems.

The value of a fire extinguisher lies in the speed with which it can be properly activated and used. For a portable fire extinguisher to be effective, several requirements must be met, including the following:

Figure 9.1 Fire extinguishers should be both easily visible and readily accessible.

- The extinguisher must be visible and readily accessible **(Figure 9.1)**.
- The extinguisher must be in working order.
- The extinguisher must be suitable for the hazard being protected.
- The person using the extinguisher must know how to operate it.

The purpose of a portable fire extinguisher is to enable an individual with minimal training and orientation to extinguish an incipient fire with reduced risk to the operator. This action must occur after notification of the fire department. As was discussed in Chapter 1, Overview of Fire Detection and Suppression Systems, there are many significant fires where notification of the building occupants and the fire department was delayed while fire extinguishers were being used. In those instances, the delay resulted in greater damage to life and property.

This chapter provides information on the basic components and types of fire extinguishers and discusses the ways in which they are classified, rated, tested, inspected, maintained, recharged, and used. Also discussed in the chapter are common extinguishing agents found in fire extinguishers as well as extinguisher selection and placement with respect to fire hazards and fire codes. NFPA® 10, *Standard for Portable Fire Extinguishers*, provides more information on the selection and use of fire extinguishers.

Extinguisher Classifications

Fire extinguishers are classified based upon the type of fire or fires they are effective at extinguishing. Fires have been broadly grouped into five classifications according to the burning characteristics of various combustible materials. These classifications include the following:

- *Class A fire* — Involves ordinary combustibles such as wood, cloth, paper, rubber, and many plastics. These fires can be extinguished by cooling, smothering, insulating, or inhibiting the chemical chain reaction.

- *Class B fire* — Involves flammable or combustible liquids and gases, including greases and similar fuels, which can be extinguished by oxygen exclusion, smothering and insulating, and inhibiting the chemical chain reaction.

- *Class C fire* — Involves energized electrical equipment, which requires the use of a nonconductive agent for protection of the extinguisher operator. If electrical power is eliminated, these fires become Class A or Class B, and may be extinguished appropriately.

- *Class D fire* — Involves combustible metals such as magnesium, potassium, sodium, titanium, and zirconium, which require the use of an agent that absorbs heat and does not react with the burning metal.

- *Class K fire* — Involves cooking oils and fats in appreciable depth. With the advent of wet chemical fire-suppression systems in commercial kitchens, manufacturers developed K-rated extinguishers. These extinguishers use an identical fire-suppression agent as the fixed system. Class K-rated agents work by forming a barrier over the product, thus smothering the fire.

Fire extinguishers are labeled according to the classification or classifications of fire that they will extinguish. It is critical that the operator be aware of the type of fire and the extinguisher type needed in order to select the proper extinguisher. Symbols are used to identify extinguishers by classification. NFPA® 10 recognizes two different methods of extinguisher recognition: pictorial system and letter-symbol system.

Pictorial System

The pictorial labeling system is the most widely used fire extinguisher identification system. The system is designed to make the selection of fire extinguishers easier through the use of picture symbols called *pictographs*. The pictographs indicate the type of fire the extinguisher is capable of extinguishing. Pictographs also indicate when *not* to use an extinguisher on certain types of fires.

If an extinguisher is suitable for use on a particular fire, the pictograph background is light blue or black. If the extinguisher is not suitable for a particular class of fire, the pictograph symbol has a black background with a diagonal red line through the symbol. Some extinguishers are used for more than one classification of fire **(Figure 9.2)**.

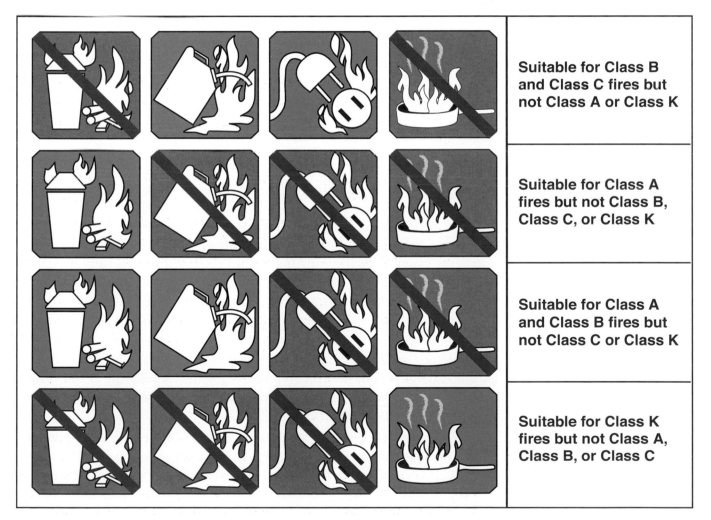

Figure 9.2 These symbols indicate when *not* to use an extinguisher on certain types of fires.

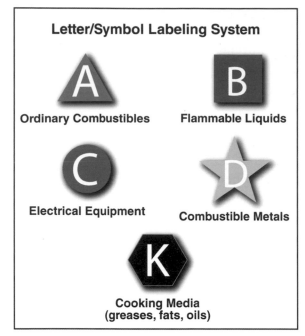

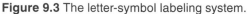

Figure 9.3 The letter-symbol labeling system.

Letter-Symbol System

The letter-symbol method of extinguisher identification is older than the pictorial system. In the letter-symbol method, each classification of fire is represented by the appropriate letter — A, B, C, D, or K **(Figure 9.3)**. A specific geometric shape encloses the letter. In addition, the background of the geometric shape can be color-coded to further identify the extinguisher. It is important to note that the letter-symbol method of fire extinguisher identification is rapidly losing its usefulness in favor of the internationally recognized pictograph system. The most recent symbol, the hexagon, is gradually being implemented by most extinguisher manufacturers for Class K applications.

Portable Fire Extinguisher Ratings

A portable fire extinguisher is rated according to its intended use and fire-fighting capability on the five classes of fire. The type and amount of agent contained in the extinguisher and the extinguisher's design determine the amount of fire that can be extinguished for a particular class of fire. This information is conveyed by an alphanumeric classification system designed by Underwriters Laboratories Inc. (UL). NFPA® 10 recommends that this rating information be displayed on the front faceplate of the extinguisher.

The rating system is based upon the extinguishment of test fires in accordance with UL 711, *Standard for Rating and Fire Testing of Fire Extinguishers.* The ratings consist of both a numeric and letter designation for extinguishers intended to combat Class A and Class B fires. Extinguishers classified for Class C fires receive only a letter rating because fires involving energized electrical equipment are fueled by materials that are typically Class A or Class B (or both) in composition. The primary concern in designating an extinguisher for Class C fires is to indicate to its user that it is safe against an electrical shock hazard. Likewise, Class D extinguishers receive only a letter designation. If an extinguisher is capable of extinguishing several classes of fire, the alphanumeric designation will denote this information by showing multiple number-letter ratings on the faceplate.

A 5- to 6-pound (2.2 kg to 2.7 kg) dry chemical extinguisher is rated at 2A or 3A depending on its chemical mixture. All other things being equal, a 4A-rated extinguisher should be able to extinguish twice as much fire as a 2A-rated extinguisher. The smallest B:C-type extinguishers have a 5B to 10B rating. All other things being equal, an extinguisher rated 40B should be twice as effective as one rated 20B.

There are several possible combinations of multipurpose extinguishers, not limited to the following:

- A:B
- A:B:C
- B:C **(Figure 9.4)**

Figure 9.4 This B:C fire extinguisher is suitable for use on both flammable liquid fires and electrical fires.

Note: Class D ratings are never assigned to any type of multipurpose extinguisher.

There are general criteria by which fire extinguishers are rated. In addition, for each specific classification rating, there are specific tests to which extinguishers must be subjected. These criteria and tests are discussed in the sections that follow.

General Rating Criteria

All extinguishers are tested in order to obtain a UL rating. In addition to the fire tests used to determine an extinguisher rating, an extinguisher must also be evaluated on the following criteria:

- Discharge volume capability
- Discharge duration
- Discharge range
- Hydrostatic testing of the vessel and discharge hose

Discharge Volume Capability

The following discharge capability characteristics are evaluated:

1. Any portable fire extinguisher that uses dry chemical or dry powder as an extinguishing agent must be able to discharge 80 percent of its contents.

2. Any portable fire extinguisher that uses agents other than dry chemical or dry powder (halon, carbon dioxide, water, etc.) must be able to discharge 95 percent of its contents.

Discharge Duration

The following discharge duration criteria must be met:

- A portable extinguisher that uses water stored under pressure as an extinguishing agent must have a minimum of a 45- to 60-second discharge time. This time period depends upon the desired rating.

- A minimum effective discharge time is required for any extinguisher for which a Class B rating is desired.

- Dry chemical extinguishers 10 pounds and larger must have a discharge rate of 1 pound per second (lb/sec).

Discharge Range

The effective range of the extinguisher stream must meet the following requirements:

1. Extinguishers that use water stored under pressure must have a minimum effective discharge of 30 feet (10 m) for a 40-second period.

2. Dry chemical and dry powder extinguishers must have a minimum horizontal discharge range of 10 feet (3 m).

Hydrostatic Testing of the Vessel and Discharge Hose

All agent storage cylinders, the discharge hose (if applicable), and the discharge nozzle are required to pass a **hydrostatic test**. This test consists of pressurizing the components to five times their rated capacity for a period of not less than 5 seconds.

Hydrostatic Test — A testing method that uses water under pressure to check the integrity of pressure vessels.

Class A Rating Tests

Class A portable fire extinguishers are rated from 1A through 40A. The ratings are based on two different test fires using various sizes of fuel cribs. These test fires include a wood-crib test and a wood-panel test **(Figure 9.5)**.

Extinguishers that are rated Class 1A through Class 6A are subjected to both tests. An extinguisher that receives a rating of Class 10A or greater is tested using only the wood-crib test. Each type of test fire is unique with respect to the configuration and amount of Class A combustibles an extinguisher must extinguish before receiving its rating.

Class B Rating Tests

Like Class A extinguishers, those used for combating Class B fires are classified with a numerical designation. Extinguishers suitable for use on Class B fires are classified with a numerical rating ranging from 1B to 640B. The number is an indication of the approximate square foot (ft^2) (m^2) area of fire involving a 2-inch (50 mm) layer of flammable liquid that can be extinguished by a novice or inexperienced operator. The rating is based on the principle that an expert extinguisher operator such as a laboratory technician can extinguish two and one-half times more fire than a novice.

For example, a novice extinguisher operator using a 60B rated extinguisher can be expected to extinguish a flammable liquid fire involving a 60 ft^2 (6 m^2) area. An expert using an extinguisher with the same rating, however, should be able to extinguish a fire involving an area of 150 ft^2 (14 m^2). Extinguishers with a rating of 20B or greater are considered suitable for outdoor fires.

Class C Rating Tests

A Class C rating is not assigned a numerical designation because the rating signifies only that the extinguishing agent is electrically nonconductive. According to UL 711, the rating is provided in conjunction with Class A, Class B, or Class K rating, such as 2A:10B:C. No effort is made to indicate the extinguisher's capacity to extinguish a fire that includes energized electrical equipment because fires involving energized electrical equipment are fueled by materials that are Class A or Class B in nature. For example, in a Class C fire, it may actually be the insulation material that is burning. The criteria for the rating of a Class C extinguisher are established by measuring the electrical conductivity of the extinguisher when it is discharged at an electrically energized target **(Figure 9.6)**.

Figure 9.5 Wood-crib tests are used to determine appropriate ratings for fire extinguishers.

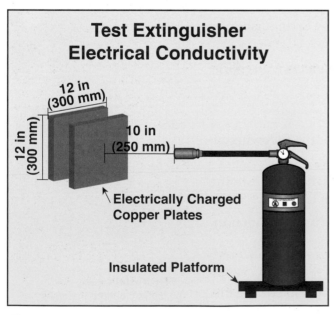

Figure 9.6 A diagram of the test layout for Class C extinguishers.

Class D Rating Tests

Class D fire extinguishers are not given numerical ratings. These types of extinguishers are generally tested against fires involving specific combustible metals including magnesium, sodium, and potassium. Other tests, which are different in nature from the standard tests, may be required to address situations or metals where the manufacturer's recommendations or the intended use of the extinguisher is indicated.

In addition to fire tests, each extinguishing agent is evaluated with respect to the adverse effects that might occur in the course of discharge. These effects include agent toxicity, fumes developed, and products of combustion. The possibility of adverse reactions resulting from the mistaken use of the agent on a combustible metal are also evaluated.

There are four magnesium fire tests that are used to evaluate Class D portable fire extinguishers, including the following:

- Area fire test
- Pallet-transfer fire test
- Premix fire test
- Casting fire test

For sodium and potassium ratings, there are two standard procedures for testing: spill-fire test and pan-fire test. Both tests are conducted when the metals are in a liquid state.

Class K Rating Tests

Class K fire extinguishers are rated for their ability to extinguish fire in commercial cooking environments. The fuel used must be new vegetable shortening or oil with an antifoaming agent and an autoignition temperature of 685°F (363°C) or higher. Tests are performed on a commercial deep fat fryer.

Extinguishing Agents

Portable fire extinguishers use many different extinguishing agents. Each agent may be able to control one or more classes of fire, but one agent will not be effective on all classes of fire. This section highlights some of the more common extinguishing agents.

Water

Water is used to extinguish fire by cooling the burning fuel. Water is inexpensive and readily available, and water extinguishers — both plain and distilled — are relatively easy to maintain **(Figure 9.7)**. However, water does have some limitations as an extinguishing agent. The use of water is not as effective as other agents on most Class B fires. Since plain water conducts electricity, it is ineffective and dangerous for use on Class C fires.

Water extinguishers have other limitations as well. They are subject to freezing and therefore should be kept in an indoor area. The weight of the water makes these types of extinguishers unwieldy, heavy to transport, and hard to maneuver. A 5-gallon (19 L) water extinguisher weighs in excess of 42 pounds (19 kg) and is the maximum size that is considered truly portable.

Figure 9.7 Water extinguishers are relatively easy to maintain.

Since distilled water has most of the minerals removed, its use is acceptable on Class C fires. The use of an atomizing applicator allows the water to be discharged in a very fine mist, which also contributes to its nonconductive characteristics.

Aqueous Film Forming Foam (AFFF)

Aqueous film forming foam (AFFF) produces both air foam and a floating film on the surface of a liquid fuel. AFFF is suitable for both Class A and Class B fires. Most commonly, the AFFF concentrate is premixed with water in the extinguisher and discharged through a special nozzle. Because the foam agent is mixed with water, AFFF is effective on Class A fires by cooling and penetrating the fuel. This agent is very effective on flammable liquid fires because of the double effect of a foam blanket and a surface film to exclude air from the fuel. The AFFF/water mixture has all of the same inherent limitations discussed for using water as an extinguishing agent.

Another type of AFFF extinguisher contains plain water in a vessel and a special nozzle with a solid form of AFFF concentrate. As the water flows through the nozzle, the concentrate is dissolved in the water to produce a **finished solution**.

> **Finished Solution —** Extinguishing agent formed by mixing foam concentrate with water and aerating the solution for expansion.

Note: Some AFFF extinguishers now include an alcohol resistance (AR) rating to identify agents that are effective on these types of fuels.

Film Forming Fluoroprotein (FFFP)

Film forming fluoroprotein (FFFP) is a foaming agent that is very similar to AFFF. This foam is usually diluted in a solution of water and 3 or 6 percent foam and is effective on Class A and Class B fires. These extinguishers are usually located where gasohol and water-soluble flammable liquids are in use.

Carbon Dioxide

Carbon dioxide (CO_2) is a colorless, noncombustible gas that is heavier than air. It extinguishes fire primarily through a smothering action by establishing a gaseous blanket between the fuel and the surrounding air. CO_2 is suitable for Class B and Class C fires. It has very limited value on Class A fires. Because of its gaseous nature, it is difficult to project it very far from the discharge horn of the extinguisher.

CO_2 is stored in the extinguisher in a liquid state, which allows more weight to be stored in a given volume. When discharged from the extinguisher, the CO_2 has a white cloudy appearance, which is due to the small dry ice crystals that may be carried in the gas stream when it is discharged **(Figure 9.8)**.

Figure 9.8 A Carbon Dioxide extinguisher in use.

Halons

Halons and other **halogenated agents** contain atoms from one of the halogen series of chemical elements: fluorine, chlorine, bromine, and iodine. The halogenated agents are principally effective on Class B and Class C fires. Halons were originally developed as *clean agents* because, they did not leave any residue when used. However, the original halogenated agents have proven to be harmful to humans and the earth's ozone layer, and restrictions

> **Halogenated Agents —** Chemical compounds (halogenated hydrocarbons) that contain carbon plus one or more elements from the halogen series. Halon 1301 and Halon 1211 are most commonly used as extinguishing agents for Class B and Class C fires.

have been placed on its production. For more information on the hazards of halogenated agents and types of agents that are still in use, refer to Chapter 8, Special Extinguishing Systems.

The main type of halon found in portable extinguishers is bromochlorodifluoromethane, commonly referred to as Halon 1211. Bromotrifluoromethane, commonly referred to as Halon 1301, is another form, but it is mainly seen in fixed extinguishing equipment.

Halon Replacement Agents

There has been considerable research and development on new clean agents that extinguish fires in the same manner as halon agents, but without the associated damage to the atmosphere. While halon replacement agents have many of the same characteristics of their halon predecessors, they are at a disadvantage due to their greater expense and need for greater quantities of agent to extinguish the fire.

There are several categories of clean agents available. There are halocarbon clean agents as well as inert gas clean agents. Halocarbons are either hydrochlorofluorocarbons (HCFCs) or hydrofluorocarbons (HFCs). For more information on halon replacement agents, refer to Chapter 8, Special Extinguishing Systems.

Dry Chemical Agents

In physical form, **dry chemical** agents are very small solid powdery particles **(Figure 9.9)**. Because these particles are solid, they can be projected more effectively from the extinguisher nozzle than gaseous agents. Therefore, dry chemicals do not dissipate into the atmosphere as readily as gases and are especially suitable for controlling outdoor fires. The primary disadvantage of using dry chemical fire extinguishers is difficulty in cleanup. An airborne dry chemical travels much farther than the immediate fire area, so cleanup can be very time-consuming.

> **Dry Chemical** — Any one of a number of powdery extinguishing agents used to extinguish fires. The most common include sodium or potassium bicarbonate, monoammonium phosphate, or potassium chloride.

As with all fire extinguishers, a dry-chemical type should never be refilled with an agent other than the specific agent for which the extinguisher was designed. Dry chemical extinguishing agents should never be mixed. This mixing can result in a dangerous chemical reaction, especially with monoammonium phosphate and other dry chemicals.

Several dry chemicals have proven useful as extinguishing agents in portable fire extinguishers. These agents, usually called *ordinary dry chemical agents*, include sodium bicarbonate, potassium bicarbonate, urea bicarbonate, and potassium chloride. In addition, there is monoammonium phosphate, which is a multipurpose dry chemical.

Sodium Bicarbonate

Sodium bicarbonate is a form of baking soda and was the first commercially produced dry chemical agent. This agent is still widely used and is treated to be water-repellent and free-flowing. When used as an extinguishing agent, sodium bicarbonate has no toxic effects but

Figure 9.9 A dry chemical extinguisher in use.

may be slightly irritating to the eyes if contact is made. Sodium bicarbonate is color-coded either with blue or white to distinguish it from other dry chemical agents. The agent is effective on Class B and Class C fires.

Sodium bicarbonate has a very rapid knockdown capability against flaming combustion and also has some effect on surface fires in Class A materials. It has been used successfully on textile machinery where fine textile fibers can produce a surface fire.

Potassium Bicarbonate

Potassium bicarbonate, which is also known as *Purple K*, is deemed twice as effective as sodium bicarbonate in fire extinguishment. It is most effective on Class B or Class C fires and is also treated to be water-repellent and free-flowing. Potassium bicarbonate is a violet color to differentiate it from other dry chemicals.

Monoammonium Phosphate

Monoammonium phosphate is an effective and popular agent for use on Class A, Class B, and Class C fires. It has an action similar to other dry chemicals and quickly knocks down flaming combustion. On Class A materials, the monoammonium phosphate forms a solid coating and extinguishes the fire by smothering.

Dry Powder Agents

Dry powder extinguishing agents are those used for extinguishing Class D fires and are not to be confused with dry chemicals **(Figure 9.10)**. *Dry powders* are designed to extinguish fires in combustible metals such as aluminum, magnesium, sodium, and potassium. Ordinary extinguishing agents are not capable of controlling fires in combustible metals. Violent reactions may occur, and toxic gases may be released when water contacts the burning metal.

A number of extinguishing agents have been developed for extinguishing fires involving combustible metals. These include some exotic and specialized materials such as foundry flux, trimethoxyboroxine, ternary eutectic chloride, and boron trifluoride. There is no single agent that is effective on all combustible metals. An extinguishing agent must be carefully chosen for the hazard being protected. Three of the more commonly encountered Class D extinguishing agents include Na-X®, Met-L-X®, and Lith-X®.

Na-X ®

Na-X® is a Class D extinguishing agent designed specifically for use on sodium, potassium, and sodium-potassium alloy fires. Na-X® is not suitable for use on magnesium fires. Chemically, Na-X® has a sodium carbonate base combined with additives to enhance flow. The extinguishing action forms a crusting or caking on the burning material, causing an oxygen-deficiency and thereby extinguishing the fire. Application can be from portable extinguishers or by scoops from pails.

Met-L-X ®

Met-L-X® is a sodium chloride-based extinguishing agent intended for use on magnesium, sodium, and potassium fires. Like other dry powders, it contains additives to enhance flowing and prevent caking in the extinguisher. It also

Figure 9.10 This dry powder extinguisher is rated for use on Class D fires.

extinguishes metal fires by forming a crust on the burning metal to exclude oxygen. The agent is applied from the extinguisher to first control the fire, and then the agent is applied more slowly to bury the fuel in a layer of powder.

Lith-X ®

Lith-X® is an agent that can be used on several combustible metals. It was developed to control fires involving lithium but can also be used to extinguish magnesium, zirconium, and sodium fires. Lith-X® consists of a graphite base that extinguishes fires by conducting heat away from the fuel after a layer of the powder has been applied to the fuel. Unlike other dry powders, this agent does not form a crust on the burning material.

Wet Chemical Agents

Wet chemical extinguishing agents are used to suppress fires involving commercial cooking equipment. Most Class K extinguishing agents are alkaline based mixtures consisting of potassium carbonate, potassium acetate, potassium citrate, or a combination. These agents are mixed with water and discharged by an expellant gas. Wet agent is emitted from the extinguisher as a fine mist, which acts to cool the flame front. The misting of the agent also helps to prevent splashing the burning oil.

The agent's primary extinguishing benefit is its ability to mix with the cooking grease to form a foam barrier, or soapy mixture, over the burning fuel through saponification. The Class K solution thus provides a blanketing effect similar to a foam extinguisher but with a greater cooling effect. The saponification process only works on animal fats and vegetable oils.

Types of Fire Extinguishers

Portable fire extinguishers use different methods to expel the extinguishing agent and can be broadly classified according to the method used. These methods include the following:

- Stored-pressure
- Cartridge-operated
- Pump-operated

Additionally, fire service personnel may encounter extinguishers that are out of production and no longer suitable for use **(Figure 9.11)**. These obsolete extinguishers should be identified and eliminated, because operation of them could lead to injury or death of the operator.

Figure 9.11 Halon fire extinguishers can still be encountered in some occupancies.

Stored-Pressure

A stored-pressure fire extinguisher contains an expellant gas and an extinguishing agent in a single chamber. The pressure of the gas forces the agent out through a siphon tube, valve, and nozzle assembly **(Figure 9.12, p. 222)**. Dry chemical extinguishers typically use nitrogen as an expellant gas. In other cases, the expellant gas can be the vapor phase of the agent itself such as that in CO_2 extinguishers. As a highly compressed gas, CO_2 forms its own expellant. Units that use a separate expelling gas have a pressure gauge that permits visual determination as to whether the extinguisher is ready to use.

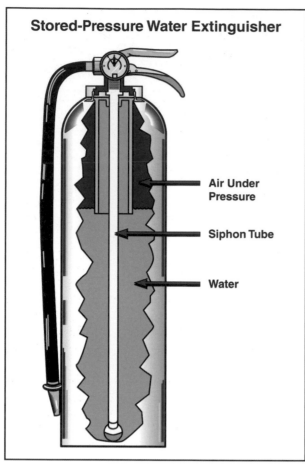

Stored-Pressure Water Extinguisher

Air Under Pressure

Siphon Tube

Water

Figure 9.12 In a stored-pressure extinguisher, the expellant gas forces the extinguishing agent out through the siphon tube.

Figure 9.13 A cross section of a cartridge-operated extinguisher.

Stored-pressure extinguishers are easy to use. They usually require only that the operator remove a safety pin and squeeze the valve handle. However, refilling the unit requires special charging equipment for pressurization, so a qualified person should perform all extinguisher servicing. Many states and municipalities require a license to perform this work. This type of extinguisher may be found in areas such as office buildings, department stores, or private residences where a high-use factor is not involved.

Cartridge-Operated

The cartridge-operated extinguisher has the expellant gas stored in a cartridge, while the extinguishing agent is contained in an adjacent cylinder called an *agent cylinder* or *tank* **(Figure 9.13)**. Upon use, the expellant, which is usually CO_2 or nitrogen gas, is released into the agent cylinder. The pressure of the gas forces the agent into the application hose. Discharge is controlled by a handheld nozzle or lever. No pressure gauge is provided.

During inspection, the gas cartridge is weighed to ensure that it has the appropriate amount of propellant. Replacing the gas cartridge and filling the agent cylinder recharges this type of extinguisher. This procedure may be performed in-house and does not require special equipment. Cartridge-operated extinguishers are found in industrial operations such as paint spraying or solvent manufacturing where they may be used frequently.

Pump-Operated

A pump-operated extinguisher discharges its agent by the manual operation of a pump. This type of extinguisher is limited to the use of water as the extinguishing agent. Its primary advantage is that it can be refilled from any available water source in the course of extinguishing a fire. Maintenance is extremely simple and consists mainly of ensuring that the extinguisher is full and has not suffered any mechanical damage.

Obsolete

Fire protection personnel must be alert for fire extinguishers that are out of production and no longer suitable for use. American manufacturers stopped making inverting-type fire extinguishers in 1969. These include soda-acid, foam, internal cartridge-operated water, loaded-stream, and internal-cartridge dry-chemical extinguishers **(Figure 9.14)**. Manufacturing of extinguishers made of copper or brass with cylinders either soft-soldered or riveted together was also discontinued at this time. Because of the toxicity of carbon tetrachloride and chlorobromomethane, extinguishers using these agents were prohibited in the workplace.

Fire service personnel will occasionally encounter obsolete fire extinguishers in old buildings. Dry chemical stored-pressure extinguishers produced prior to 1984 must be removed from service at the next maintenance interval.

In addition, if any fire extinguisher cannot be maintained per manufacturer's specifications, it must be removed from service. These extinguishers should be disposed of according to the operating procedures of the authority having jurisdiction (AHJ).

Selection and Distribution of Extinguishers

Extinguishers must be properly distributed throughout a facility to ensure that they are readily available during an emergency. To be effective, extinguishers must be located in close proximity to where they may be needed. In the same manner, extinguishers cannot be effective if there are not enough of them provided for the hazard involved.

Requirements for extinguisher distribution are found in NFPA® 10. These requirements are separated into specifics for Class A, Class B, Class C, Class D, and Class K hazards. Because local codes and ordinances can be more restrictive, these should be reviewed along with the requirements contained in NFPA® 10. Important elements in the selection and distribution of fire extinguishers include the following:

- Chemical and physical characteristics of the combustibles that might be ignited

- Potential severity, size, intensity, and rate of advancement of fire

- Location of the extinguisher

- Effectiveness of the extinguisher for the hazard in question

- Personnel available to operate the extinguisher, including their physical abilities and any training they may have in the use of extinguishers

- Environmental conditions that may affect the use of the extinguisher such as temperature, winds, and the presence of toxic fumes or gases

- Any anticipated adverse chemical reactions between the extinguishing agent and the burning material **(Figure 9.15)**

- Any health and occupational safety concerns such as exposure of the extinguisher operator to heat and products of combustion during fire-fighting efforts

- Inspection and service required to maintain the extinguishers

In addition, the type, size, and number of extinguishers needed may vary according to the type of occupancy and the nature of the hazard. NFPA® 10 discusses extinguisher selection based upon the nature of the hazard and the size of the extinguisher.

A general method found in NFPA® 10 is used to determine a satisfactory distribution of extinguishers in the vast majority of situations. This method classifies occupancies as light or low hazard, ordinary or moderate hazard, or extra or high hazard. Fire extinguisher distribution is specified on the basis of that classification.

Figure 9.14 Soda-acid extinguishers are obsolete but may still be encountered in some occupancies.

Figure 9.15 Extinguishers must be tailored to the hazard. In facilities that contain combustible metals, Class D extinguishers must be supplied.

In addition, NFPA® 10 recommends the minimum size of extinguisher and the maximum area to be protected by an extinguisher. Tables contained in NFPA® 10 provide information for the computation of square footage (m²) of a facility coupled with the size of extinguisher required for a minimum number of extinguishers that must be installed. These tables also take into account the hazard classification of the facility.

Class A Factors

In ordinary- or low-hazard occupancies, the AHJ may approve the use of several lower-rated extinguishers in place of one higher-rated extinguisher. For example, two or more extinguishers may be used to satisfy a 6A rating if there are enough individuals trained to use the extinguishers. When the weight of the extinguisher causes problems for those who will be operating it, two extinguishers of lesser weight may be used to replace the heavier extinguisher.

Class B Factors

Determining the distribution of Class B extinguishers depends upon the travel distance to the hazard. Flammable liquid fires develop very rapidly and occur in a variety of situations that are fundamentally unique from a fire-control standpoint. In providing extinguishers for Class B hazards, two situations may be encountered: One is a spill fire where the flammable liquid does not have depth. The other involves a flammable liquids fire where the liquid has depth such as with dip tanks. NFPA® 10 establishes ¼ inch (6 mm) as the criterion for a flammable liquid fire to be classified as having depth. Anything less is considered to be without depth.

Class C and Class D Factors

There are no special spacing rules for Class C hazards because fires involving energized electrical equipment usually involve Class A or Class B fuels. Furthermore, the placement and distribution of fire extinguishers for Class D combustible metals cannot be generalized. Determining extinguisher placement involves making an analysis of the specific metal, determining the amount of metal present, determining whether the metal is solid or particulate, and knowing the characteristics of the extinguishing agent. NFPA® 10 recommends that the travel distance for Class D extinguishers not exceed 75 feet (23 m).

Class K Factors

In the working environment of commercial cooking occupancies, the potential for a hostile fire is always present. Employees in such areas are charged with the responsibility to maintain appropriate cooking temperatures to ensure safety. Because employees are in various levels of training for their jobs and because there is the potential of fire hazards occurring in an assembly area such as a dining room, NFPA® 10 assigns a more restrictive distance requirement. In areas where Class K fires are likely, the maximum travel distance from the hazard to the extinguisher is 30 feet (9 m) **(Figure 9.16)**.

Installation and Placement of Extinguishers

In addition to proper selection and distribution, the effectiveness of fire extinguishers requires that they be readily visible and accessible. Proper extinguisher placement is an essential, but often overlooked aspect of fire

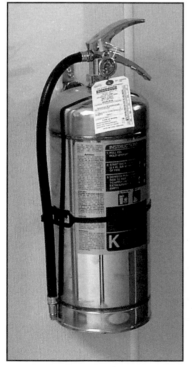

Figure 9.16 Class K extinguishers should be located no more than 30 feet from the hazard.

protection. Extinguishers should be mounted properly to avoid injury to building occupants and avoid damage to the extinguisher. Some examples of improper mounting would be an extinguisher mounted where it protrudes into a path of travel or one that is sitting on top of a workbench with no mount at all. To minimize these problems, extinguishers are frequently placed in cabinets or wall recesses for protection of both the extinguisher and the people who might walk into them **(Figure 9.17)**. If an extinguisher cabinet is placed in a rated wall, then the cabinet must have the same fire rating as the wall assembly. In properly placing fire extinguishers, the following factors must be taken into account:

- Visible and well-signed
- Not blocked by storage or equipment
- Near points of egress or ingress
- Near normal paths of travel

Although it is critical that an extinguisher be properly mounted, it must be placed so that all personnel can access it. For safe lifting, the extinguisher should not be placed too high above the floor. The standard mounting heights specified for extinguishers are as follows **(Figure 9.18)**:

- Extinguishers with a gross weight of 40 pounds (18 kg) or less should be installed so that the top of the extinguisher is not more than 5 feet (1.5 m) above the floor.
- Extinguishers with a gross weight exceeding 40 pounds (18 kg), except wheeled types, should be installed so that the top of the extinguisher is not more than 3½ feet (1 m) above the floor.
- The clearance between the bottom of the extinguisher and the floor should never be less than 4 inches (100 mm).

Figure 9.17 Extinguishers should be properly mounted to protect both the extinguisher and occupants.

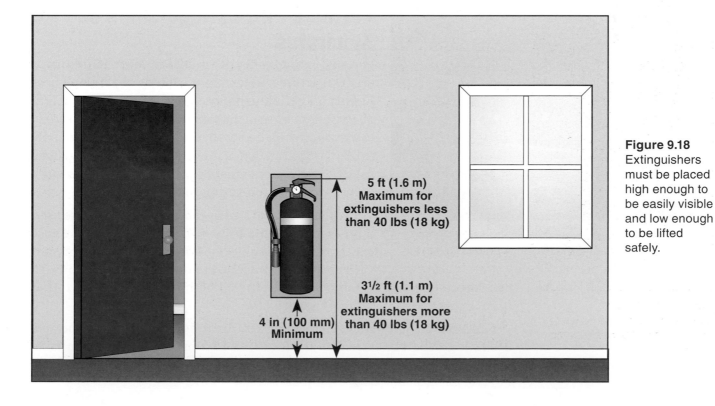

5 ft (1.6 m)
Maximum for extinguishers less than 40 lbs (18 kg)

3½ ft (1.1 m)
Maximum for extinguishers more than 40 lbs (18 kg)

4 in (100 mm)
Minimum

Figure 9.18 Extinguishers must be placed high enough to be easily visible and low enough to be lifted safely.

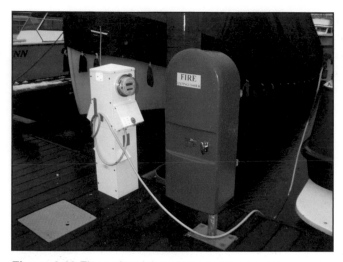

Figure 9.19 Fire extinguishers intended for marine applications are typically placed in cabinets to protect the extinguisher from corrosion. *Courtesy of Gaylen Brevik, Castle Rock Firefighting Consulting.*

The physical environment is very important to an extinguisher's reliability. Of greatest concern is the temperature of the environment. Because testing laboratories evaluate water-based extinguishers at temperatures between 40°F and 120°F (4°C and 49°C), these extinguishers must be located where freezing is not possible. Other types of extinguishers can be installed where the temperature is as low as -40°F (-40°C). Specialized extinguishers are available for temperatures as low as -65°F (-54°C).

Extinguishers using plain water can be provided with antifreeze recommended by the manufacturer. However, care must be exercised in the use of antifreeze. Ethylene glycol should not be used in portable fire extinguishers. Calcium chloride cannot be used in stainless steel units. Antifreeze cannot be added to AFFF extinguishers.

Other environmental factors that may adversely affect an extinguisher's effectiveness are snow, rain, and corrosive fumes. A corrosive atmosphere can be encountered not only in an industrial environment but also in marine applications where extinguishers are exposed to saltwater spray. For marine applications, extinguishers are available that have been listed for use in a saltwater environment. In the case of outdoor installations, the extinguisher can be protected with a plastic bag or placed in a cabinet **(Figure 9.19)**.

Tampering is also an issue that can affect an extinguisher's ability to operate. Locations such as college campuses and public facilities are especially vulnerable to vandalism. Products are commercially available to limit tampering with or removal of extinguishers in these locations.

Figure 9.20 Extinguishers on fire apparatus should be secured in compartments instead of being exposed to the elements.

Portable Fire Extinguishers on Fire Apparatus

Portable extinguisher use is not limited to buildings or other structures. In fact, in the hands of a trained firefighter, the extinguisher probably attains its maximum effectiveness. Most of the same considerations that apply to the use of extinguishers by the general public also apply to their use by fire department personnel. Extinguishers used by firefighters must be properly selected, readily accessible, and properly maintained.

Extinguishers carried by the fire department are used more frequently than those in private industry. These extinguishers are subject to harsher use and to a greater variety of environmental conditions and vibrations. To be protected, extinguishers should be carried within compartments rather than in an exposed location **(Figure 9.20)**.

NFPA® 1901, *Standard for Automotive Fire Apparatus,* requires that portable fire extinguishers be carried on apparatus. The extinguishers must be suitable for Class A, Class B, and Class C fires. To date, the NFPA® does not require a

Class K extinguisher on fire apparatus. Although individual fire extinguishers are specified by the apparatus purchaser, minimum ratings for the different types of extinguishers to be carried on fire apparatus are as follows:

1. One dry chemical extinguisher rated for B:C fires with at least an 80-B:C rating.

2. One 2½ gallon or larger water extinguisher.

Inspection, Maintenance, and Recharging

Proper servicing is essential to maintaining portable fire extinguisher readiness. The following section highlights the procedures required for properly inspecting, maintaining, and recharging portable fire extinguishers.

Inspection

An inspection is a quick check that visually determines whether a fire extinguisher is properly placed and operable. A regular inspection ensures that the extinguisher is fully charged and ready to use **(Figure 9.21)**. Inspections verify that the extinguisher is properly located, is not blocked in any way, and has not been tampered with or emptied. Since extinguishers in most occupancies are used infrequently, this regular inspection is very important to ensure their state of readiness. While this may seem a trivial function, it is critically important because one extinguisher that does not properly operate can result in significant property loss or personal injury.

Figure 9.21 A regular inspection ensures that the extinguisher is fully charged and ready to use.

In order to be effective, inspections must be conducted frequently. NFPA® 10 recommends that extinguisher inspections be performed monthly. It is also important that complete records of the inspections are maintained. Inspection tags can be used for recording the inspector's name and the date of the inspection. Items that should be checked during a monthly inspection include the following:

- Proper location
- Access (visible and accessible)
- Inspection tag (check for annual inspection)
- Horn or nozzle (look for obstructions)
- Lock pins and tamper seals (ensure they are intact) **(Figure 9.22)**
- Signs of physical damage
- Pressure gauge (ensure extinguisher is fully charged)
- Applicability of extinguisher for hazard classification

Figure 9.22 Extinguishers should be checked to ensure that lock pins and tamper seals are intact.

Maintenance

The purpose of extinguisher maintenance is to ensure that the extinguisher will operate safely. Maintenance must be performed by a qualified service technician employed by an extinguisher distribution and/or service company.

Maintenance involves a thorough examination of the mechanical parts, the extinguishing agent, and the expelling means. Maintenance should be performed periodically as required by the AHJ, after each use, or when an inspection shows obvious damage.

Recharging

Recharging is the replacement of the expellant and, if necessary, the agent in a fire extinguisher. This is one of the most critical procedures in the maintenance of extinguishers. Recharging is not required on a periodic basis for every type of extinguisher. For more information on fire extinguisher maintenance intervals, see NFPA® 10.

Figure 9.23 Only qualified personnel should refill and recharge fire extinguishers.

Recharging some extinguishers involves not only filling the unit with the proper agent, but pressurizing it as well. Pressurization must be performed by those properly trained and with the proper equipment. One obvious danger in pressuring an extinguisher is applying too much pressure to the cylinder. It is important, therefore, to use a source of expellant gas with a pressure not greater than 25 psi (170 kPa) above operating pressure. Another potential danger is the inclusion of moisture in nonwater extinguishers. Not only can moisture result in the caking of dry chemical, but it also can contribute to interior corrosion and eventual failure of the container.

When refilling an extinguisher, it is important that only those chemicals or materials specified by the manufacturer or those having an equivalent composition be used. For example, dry chemicals must be of proper particle size to flow properly. Although sodium bicarbonate is one of the dry chemicals used in extinguishers, the baking soda purchased in a grocery store is not suitable for use in an extinguisher. Extinguishing agents cannot be mixed, and only the extinguishing agent intended for a particular extinguisher can be used in that extinguisher.

Only qualified personnel should refill and recharge any type of fire extinguisher **(Figure 9.23)**. Even the cartridge-operated type, which is the easiest to refill, should be serviced by properly trained technicians.

Hydrostatic Testing

With the exception of pump-tank water extinguishers, fire extinguishers are considered pressure vessels. They are either maintained under a constant pressure or are pressurized when they are used. The pressure in extinguishers varies from 100 to 850 psi (689 kPa to 5 860 kPa) depending upon the type. Physical damage and corrosion can cause extinguisher shell failure, which may result in severe injury or death. To ensure that an extinguisher is strong enough to withstand the pressures to which it is subjected, it must be periodically tested.

The method used to pressure test an extinguisher is known as the *hydrostatic test*. Hydrostatic testing consists of filling the cylinder with water and then applying the appropriate pressure by means of a pump. This method is used because it is safer than using a compressed gas. If the cylinder fails while pressurized with water, a violent rupture typically does not occur. Hydrostatic testing of extinguishers is performed at the intervals specified in NFPA® 10 or whenever there is evidence of damage or corrosion. Hydrostatic testing should only be performed by trained and experienced personnel.

The hydrostatic test pressure is based on the extinguisher's service pressure and its factory test pressure. For CO_2 extinguishers and CO_2 or nitrogen cylinders used as an expellant source, the hydrostatic test pressure is 5/3 or 167 percent of the service pressure stamped on the vessel. The factory test pressure is the pressure at which the vessel was tested at the time of its manufacture. This pressure is shown on the extinguisher nameplate **(Figure 9.24)**. For Halon 1211 and stored-pressure extinguishers, the hydrostatic test pressure is the factory test pressure.

Figure 9.24 The factory test pressure is shown on the extinguisher nameplate.

An extinguisher should not be subjected to hydrostatic testing and should be removed from service if it shows signs of physical defects such as damaged threads, corrosion, or welded repairs **(Figure 9.25)**. If an extinguisher vessel is of soldered or riveted brass or copper construction and has been burned in a fire, it should not be hydrostatically tested. Extinguisher vessels with a stainless steel shell that have contained calcium chloride and have been burned in a fire should also not be hydrostatically tested. If an extinguisher vessel ever fails a hydrostatic test, it must be removed from service, made inoperable, and properly destroyed.

As with inspections, records of extinguisher maintenance and hydrostatic testing are important parts of a fire protection program. Extinguisher maintenance that included disassembly is recorded on a plastic collar tag that is secured to the neck of the extinguisher handle or other hardware. The information on this tag includes the month and year that the maintenance was performed and the name of the person who performed the work. Hydrostatic test results are recorded on the shell of the extinguisher. Along with the labels attached to the extinguisher, a record-keeping system should be established for management purposes.

Portable Fire Extinguisher Use

The effectiveness of an extinguisher in dousing a hostile fire depends upon several factors. The most important factor is the user. One person may be able to extinguish a fire while another individual, given the same fire, may not **(Figure 9.26)**. Many extinguishers completely discharge the extinguishing agent in 8 to 15 seconds, which leaves little time for trial and error or practice. Improper use of an extinguisher may injure the operator and may also delay extinguishment.

Figure 9.25 Extinguishers should not be subjected to hydrostatic testing if they show signs of damage.

Figure 9.26 A firefighter would be expected to extinguish much more fire with an extinguisher than a bystander because of training and experience.

Figure 9.27 It is necessary to pull the pin on an extinguisher before it can be operated.

Figure 9.28 A test burst from the extinguisher should be done prior to approaching the fire to ensure the extinguisher is operational.

Upon discovery of any fire, the first action should be the activation of the alarm and notification of the fire agency. In a commercial or industrial occupancy, this action might mean pulling the fire alarm or using a voice paging system to activate an emergency response team. In a smaller environment, such as a residence or small office, this certainly means calling the emergency number, such as 9-1-1, to summon the fire agency. It is critical that this action take place immediately because valuable time is lost when fighting a fire unsuccessfully.

It is important that an extinguisher operator choose the appropriate extinguisher for the fire. The picture-symbol and letter-symbol methods of identification provide a quick means of matching an extinguisher to a classification of fire. This identification is one step that should not be overlooked because choosing the wrong extinguisher could have devastating effects. In many situations, this will not be an issue because extinguishers have been appropriately selected initially for the hazards present in the area or facility. However, individuals should always be encouraged to verify that they have the proper extinguisher for the fire before using it.

The use of a fire extinguisher is designed for an untrained person and usually involves three to four basic steps. There are general steps that apply to the use of all extinguishers. These steps are generally referred to as the *PASS method*. There are also some specific instructional techniques that apply to the use of extinguishers on specific classifications of fire.

PASS Method

Extinguisher operation has been simplified so that even untrained personnel can use them effectively. The acronym PASS serves as a memory aid for the proper operating steps. It is useful as a teaching tool for either firefighters or the general public. During a fire, every second is of great importance. Therefore, everyone should be acquainted with operating instructions. Steps for the PASS method of extinguishment are as follows:

- *P* — Pull the pin at the top of the extinguisher **(Figure 9.27)**. Break the plastic or thin wire inspection band as the pin is pulled. For cartridge-operated extinguishers, remove the nozzle from its holder. Pressurize the extinguisher by depressing a lever that punctures the cartridge seal.

- *A* — Aim the nozzle or outlet toward the fire. Note that some hose assemblies are clipped to the extinguisher body. The nozzle should be pulled away from the extinguisher and pointed toward the fire.

- *S* — Squeeze the handle above the carrying handle to discharge the agent. Give a short test burst to ensure that the extinguisher is operational before approaching the fire **(Figure 9.28)**. Release the handle to stop the discharge of the agent at any time (operates similarly to an aerosol can).

- *S* — Sweep the nozzle back and forth at the base of the fire to disperse the extinguisher agent **(Figure 9.29)**. Watch for remaining smoldering hot spots or possible reignition of flammable liquids. Ensure that the fire is completely extinguished.

Extinguisher operators should always maintain a means of exit from the fire area and should be prepared to retreat if conditions deteriorate **(Figure 9.30)**. While the PASS method is appropriate for most extinguishers, some differences exist in the fire attack depending on the extinguishing agent and classification of fire.

Class A Fire Attack

When using a water-based extinguisher, the stream must be aimed at the seat of the fire to maximize the cooling effect of the water on the fuel. Initially, the extinguisher should be used at a distance of 10 to 30 feet (3 m to 9 m) from the fire. Attacking a fire from more than 30 feet (9 m) will not be very effective **(Figure 9.31)**.

When the flames are knocked down, the operator should move closer to wet down any remaining smoldering materials. Fires involving compacted fuels or other deep-seated burning materials must be thoroughly soaked and pulled apart to reach the remaining fire. The extinguisher can be used intermittently to facilitate the soaking. If possible, the material should be moved outside to complete the overhaul process. A thumb or finger can be placed partially over the nozzle orifice to break the stream into a spray pattern if desired **(Figure 9.32)**.

AFFF and other foam-type extinguishers are effective against Class A fires and are used in a manner similar to water extinguishers. These agents have a low surface tension, which allows the agent to penetrate the fuels, especially tightly packaged fuels.

Figure 9.29 The nozzle should be swept back and forth to adequately distribute the extinguishing agent.

Figure 9.30 Firefighters should be trained to always maintain a means of exit when using a fire extinguisher.

Figure 9.31 A Class A fire should be attacked at a distance of 10 to 30 feet depending on the reach of the extinguisher.

Figure 9.32 A finger placed over the nozzle orifice can break up the stream pattern if necessary.

When using a multipurpose dry chemical extinguisher on a Class A fire, the fire should be attacked at its base while sweeping the nozzle from side to side. Because the multipurpose agent forms a coating on the fuel, it is important that the dry chemical agent thoroughly coats all fuel surfaces. The dry chemical has little cooling effect, and deep-seated fires are difficult to extinguish. If the fire involves lightweight materials that might scatter easily, the dry chemical extinguisher can be used approximately 10 feet (3 m) from and 3 feet (1 m) above the fire. This procedure will cause a cloud of agent to form over the fire and the surrounding area.

Figure 9.33 Attacking a Class B fire with a fire extinguisher.

Figure 9.34 AFFF extinguishers are effective on Class B fires.

Hydrocarbon — Organic compound containing only hydrogen and carbon and found primarily in petroleum products and coal.

Class B Fire Attack

Both regular and multipurpose dry chemical extinguishers are used to extinguish fires involving flammable or combustible liquids and gases **(Figure 9.33)**. The extinguishing agent should initially be discharged from a distance of approximately 10 feet (3 m). If the attack is started at a closer range, the velocity of the dry chemical discharge may cause the fuel to splash, thus spreading the fire. In general, a flammable liquid fire should be attacked by sweeping the leading edge of the fire with the agent and moving forward to continue applying the agent. This action interrupts the chemical chain reaction and reduces the radiant heat. When fighting flammable liquid fires, it is possible for the fire to flash back across the surface of the liquid. This situation occurs because the extinguishing action of the dry chemical — the interruption of the chemical reaction — is not cumulative like it is with foam. Therefore, an operator must stay alert and be prepared for immediate retreat if a flashback occurs.

If it is necessary to attempt a second attack on the fire, it must be done with a full extinguisher. Flammable liquid fires have the characteristic of being particularly fast spreading. If any doubt exists about the ability to control the fire, personnel should leave the area and await the arrival of the fire department.

CO_2 extinguishers are also effective on flammable liquid fires. Because it is a gas, CO_2 cannot be projected very far out of the nozzle. Therefore, it must be applied at a closer range than other agents. The agent should be applied by sweeping it across the surface to overlap the burning liquid. CO_2 has both a cooling and smothering effect. Consequently, discharge should be continued after initial extinguishment to cool the fuel and prevent flashback.

AFFF is effective on Class B fires involving **hydrocarbons** such as fuel oils, gasoline, and kerosene **(Figure 9.34)**. Without an AR rating, AFFF is not effective on flammable liquids like acetone, alcohols, ethers, and lacquer thinners because they dissolve the foam. AFFF is also not effective on pressurized liquid and gases.

Because AFFF extinguishes fire by establishing a barrier that excludes air and oxygen, the application technique must enhance the formation of the surface film. Maintenance of the foam blanket should be made a priority after extinguishment to prevent reignition. AFFF should not be discharged directly into the liquid surface because it will penetrate the burning liquid and cause splashing. On fires of depth, the foam should be deflected off the sides or back of the enclosing tank so that the agent will flow down on the liquid. On spill fires, the AFFF can be directed onto the surface just in front of the fire to spray over the fire.

Class C Fire Attack

When attacking fires involving energized electrical equipment, the primary consideration is to ensure that the agent is dielectric or electrically nonconductive so that the operator is not injured. If possible, the equipment should be de-energized before initiating the attack. Whether de-energized or not, fire involving electrical equipment can be fought effectively with halogenated agent, CO_2, dry chemical, or even a water-mist extinguisher. Originally designed to take the place of halon-type fire extinguishers, the distilled water fire extinguisher has also proven to be effective in extinguishing Class C fires.

CO_2 or water-mist extinguishers are the best choices for use on sensitive electrical equipment. Both agents are nonconductive, noncorrosive, and leave no residue. CO_2 should be applied at close range for a quick knockdown. Water-mist stored-pressure extinguishers can be used from a somewhat greater distance, even as far as 12 feet (3.5 m). The application wand on a water-mist-type extinguisher is designed for operator safety. Halon is effective on Class C fires but is toxic to both the operator and the environment.

Class D Fire Attack

Various dry powder agents will extinguish fires involving magnesium, sodium, and potassium alloys. Each agent has its limitations, however, and these must be known before attacking a Class D fire. For example, even though a particular agent may work well in combating lithium fires, it may not be effective on magnesium fires. Most dry powder agents extinguish combustible metals by caking and adhering to the material, which excludes the necessary air required for combustion. The burning material should be covered with a 2-inch (50 mm) layer of dry powder agent. Dry powder agents should be applied with a minimum of disturbance to the burning material. Most are applied with either an extinguisher or a shovel **(Figure 9.35)**.

Figure 9.35 Class D extinguishing agents are commonly applied with a shovel.

Class K Fire Attack

Class K fires are particularly difficult to extinguish because of their tendency to reignite. Although a Class K fire may have been extinguished properly with a dry chemical, the fuel can change chemically and reach autoreignition at a lower temperature. For this reason, only an extinguisher with a Class K rating is recommended for use on this type of fire. A Class A:B:C- or Class B:C-rated fire extinguisher may indeed work effectively, but the operator should pay attention for a possible autoreignition and a second or third attack on the fire.

The initial attack on a fire with a Class K extinguisher is similar to other types of attack. Application of the agent should begin from a distance of 10 to 12 feet (3 m to 3.5 m) from the burning material. The operator should hold the application wand at the edge of the flames and then coat the surface of the material with a side-to-side sweep. This application should continue until the fire extinguisher is completely empty. Fire extinguishment takes place through the removal of oxygen and a partial cooling of the fuel. It is the innate cooling quality of the agent, along with its ability to form soapy foam or *saponify*, that prevents the fuel from reaching a lower reignition temperature.

Summary

Portable fire extinguishers have long been recognized and used as a first line of defense against incipient fires. When extinguishers are correctly selected and placed, inspected, maintained, and used with proper training, they have proven to save lives and property.

Not all hazards are alike and therefore not all portable extinguishers are alike. In addition, extinguishers are useless if they are unavailable, not suited for the particular fire, or nonfunctioning. Personnel responsible for selection and placement of the correct extinguisher in a facility are critical components to fire and life safety. These personnel require knowledge of the operating principles of various extinguishers, the advantages and disadvantages of different extinguishing agents, and familiarity with inspection and maintenance procedures. Personnel must also be alert for obsolete or damaged extinguishers so that they can be removed from service before they cause injury when someone attempts to use them. Finally, individuals in a workplace need to be trained in the basics of extinguisher operation so that they may be able to control a fire and evacuate the premises until additional help arrives.

Review Questions

1. What are the classifications of portable fire extinguishers?
2. What is the symbol system most widely used on portable fire extinguishers?
3. What are the ratings of portable fire extinguishers based upon?
4. What types of agents are used in portable fire extinguishers?
5. What are the methods used to expel the extinguishing agent in portable fire extinguishers?
6. What standard contains information on the selection and distribution of portable fire extinguishers?
7. What are factors to consider in properly placing portable fire extinguishers?
8. What are items to check during a monthly inspection of portable fire extinguishers?
9. What is hydrostatic testing?
10. What are the steps in the PASS method?

Appendices

Contents

Divider page photo courtesy of Los Angeles Fire Department - ISTS

Appendix A
Fire Pump Testing

There are two categories of pump tests that must be performed on a stationary fire pump. The first is the initial or acceptance test, which is performed by the pump manufacturer, the engine manufacturer, the controller manufacturer, or other factory-authorized representatives. Much the same as testing a fire suppression pump mounted on a pumping apparatus, the acceptance test is conducted prior to the ownership of the pump being transferred to the end user.

Prior to the acceptance test, all local authorities should be notified of the testing time and location and invited to participate. During the test, all peripheral equipment that affects the operational capabilities of the pump is to be included in the test. The multiple control interwiring, the pressure maintenance pump, and any alternate power supply systems are to be completed, checked, and approved by the electrical contractor before testing begins.

The second category of stationary fire pump tests includes those of a periodic nature: weekly, monthly, quarterly, semi-annually, or annually. Usually associated with normal maintenance inspections, these tests are intended to ensure the continued acceptable performance of the fire pump and its associated systems. Periodic testing and maintenance is normally the responsibility of the owner/operator of the facility where the pump is installed. Occasionally, through maintenance agreements, insurance requirements, or other pump maintenance and performance contracts, the original manufacturer may retain responsibility for the continued performance of the pump. It is incumbent upon the local fire authority to clarify these issues before granting approval of and for the fire pump installation.

Acceptance Tests

The primary performance criteria for standard fire pumps are contained in NFPA® 20, *Standard for Installation of Stationary Pumps for Fire Protection*. Additionally, FM Global and other nationally recognized testing laboratories provide Pump Acceptance Test Data check sheets that also contain information on industrial fire pumps. The material contained in these checklists is compiled from the requirements of NFPA® 20, which allows the inspector to systematically check the components and performance of the fire pump. Prior to the installation of the pump and before any acceptance testing, the manufacturer must provide certified shop (pump) test characteristic curves showing head capacity and the brake horsepower of the fire pump. This is used to compare the expected performance with the fire pump's actual performance during the test. Once installed, a new industrial fire pump must be capable of satisfying three test points.

- The first point for the vertical-shaft pump is the maximum point of 140 percent of rated pressure at shutoff.
- The second point is a minimum of 100 percent of rated capacity at 100 percent of rated pressure.
- The third point specifies 65 percent of rated pressure while delivering 150 percent of the rated number of gpm.

Another component of a field acceptance test includes the flushing and hydrostatic testing of all suction and discharge piping associated with the operation of the fire pump. The installing contractor is required to furnish a certificate indicating that flushing and the hydrostatic testing have been completed. Many local authorities issue permits and successful test certificates after they have witnessed these procedures. Check with the local authority having jurisdiction (AHJ) for their requirements.

Periodic or Routine Fire Pump Maintenance and Testing

All fire pumps must be inspected, tested, and have routine maintenance performed on a regular basis as described in NFPA® 25, *Standard for the Inspection, Testing, and Maintenance of Water-Based Fire Protection Systems*. Although maintenance and testing are performed as often as the AHJ requires, the standard recognizes that a minimum interval for each of these must be adopted. Weekly tests are more cursory than annual tests and usually involve the activation of the fire pump system while recording pressure gauge information. Annual tests are far more comprehensive. These tests require actual fire flow and the determination of continued satisfactory performance of the fire pump and its associated systems during a calibrated testing format. Every test must be recorded and retained in the permanent inspection and maintenance files for the fire pump.

Weekly Tests

Weekly inspection should address four main areas of concern. These areas include the following:

- General housekeeping conditions
- The condition of the pump and the water supply system
- The electrical system
- The power system (driver) that drives the pump

The general condition of the pump room is inspected primarily to ensure adequate heating of the pump room space to prevent the pump and piping from freezing. Additionally, the proper operation of ventilation louvers must be verified so that air for combustion is in abundant supply as well as air for cooling of the motor and fire pump.

All pump system components must be visually inspected for obvious signs of malfunction or damage including pipe/seal leaks, both water and oil. System pressure gauges must be within system operational parameters. The oil in any priming or suction reservoirs must be checked and replenished if it is below manufacturer's recommendations.

The electrical systems must also be in full operating condition. The pilot light on the controller indicating that power is on must be lit as well as that of the transfer switch. The isolating switch is to be closed with the emergency power standby source ready and available. The electrical phase alarm pilot light is to be off or the normal phase rotation light illuminated (three-phase electrical motors can be seriously damaged or destroyed if an "out of phase" electrical current is supplied). Additionally, if one leg of the current is absent or of lower-than-expected current the motor will overheat and burn up.

For diesel engine driver systems, the fuel tank level must be at least two-thirds full. The controller switch is to be in the AUTOMATIC position. All batteries are to be fully charged and all current gauge readings within normal ranges. No alarm pilots may be active. If cooling water is used, the reservoir is to be full with no fluid leaks.

Weekly pump tests are also required by NFPA® 25. These tests are conducted by automatically starting the fire pump but not by flowing water. Electric pumps are run for 10 minutes and diesel pumps for a minimum of 30 minutes. These tests are to be conducted in the presence of qualified personnel who can make adjustments if any deficiencies are noted. The required performance checks include the pump system where pressure gauge readings are taken and the pump starting pressure is recorded. Pump packings are checked for leakage and adjusted if necessary. Unusual vibrations and noises during the operation are checked and adjustments are made to eliminate them. The packing boxes, bearings, and pump casing are monitored for overheating.

When testing a diesel engine system, check and record the time for the engine to crank and the time it takes for the engine to reach its running speed. The engine oil pressure gauge and the speed, water, and oil temperature indicators are to be monitored while the engine is running. A steam system is to have the steam pressure checked and recorded as well as the observed time for the turbine to reach running speed.

Annual Tests

Annual fire pump assembly tests are to be conducted to ensure the performance of the pump at minimum, rated, and peak flows. These tests are performed by flowing and measuring the pump suction, discharge pressure, and each hose stream supplied by the pump. The pressure relief valve must be monitored to ensure that the design pressure of the system is not exceeded and is observed during each flow condition (no-flow, flow) to determine if it operates at the proper pressure.

For fire pumps that feature an automatic transfer switch, several additional tests are necessary to ensure that overcurrent protective devices do not open. To conduct this test, a simulated power failure condition occurs when the pump reaches peak load. The transfer switch should operate, shifting to the alternate power source. A successful test will be recorded once it is verified that the pump continues to operate at peak load. Once this has been successfully completed, the power failure condition is removed. After the designated time delay the pump is reconnected to the normal power supply system.

It is *not* acceptable to increase the engine speed past the rated speed of the pump to reach rated pump performance. The results of the tests must be evaluated by qualified individuals. The pump assembly will be given an acceptable test performance evaluation if the test matches the initial unadjusted field acceptance test curve of the pump or if the fire pump matches the performance characteristics listed on the pump nameplate.

NOTE: For an in-depth description of the weekly and annual testing procedures for a fire pump, refer to NFPA® 25, *Standard for the Inspection, Testing, and Maintenance of Water-Based Fire Protection Systems*.

Appendix B
Fire Department Operations at Sprinklered Occupancies

Firefighters' knowledge of sprinkler system operation enhances fire fighting operations in buildings protected by sprinkler systems. At the very least, sprinklers can save fire fighting units a great deal of time and effort. Proper fire fighting tactics in sprinklered properties should proceed in support of an operating sprinkler system. Fire departments should adopt standard operating procedures (SOPs) for preincident planning and operations at sprinklered occupancies. Refer to the following NFPA® standards for additional information about sprinklered occupancies and dwellings equipped with these systems:

- NFPA® 13, *Standard for the Installation of Sprinkler Systems*

- NFPA® 13D, *Standard for the Installation of Sprinkler Systems in One- and Two-Family Dwellings and Manufactured Homes*

- NFPA® 13R, *Standard for the Installation of Sprinkler Systems in Residential Occupancies up to and Including Four Stories in Height*

Preincident Planning at Sprinklered Occupancies

Fire department SOPs cannot be established unless fire department personnel are familiar with the sprinklered properties in their area of jurisdiction. Inspections and preincident planning visits are very useful in developing information necessary to develop SOPs. Information that should be noted about the sprinkler system includes the following:

- Buildings and nature of occupancies protected by automatic sprinklers

- Type of sprinkler system(s)

- Water supply to the sprinklers, including available volume and pressure

- Locations of sprinkler valves (interior and exterior) and the area each valve controls

- Locations of fire department connections, the specific area each serves, and the nearest water supply that will not interfere with the supply to the sprinkler system(s) and the location of the nearest fire hydrant

- Pumper having primary responsibility for charging the fire department connection

- Alternate means for supplying water to the system in case of damage to the fire department connection, such as the test header at a fire pump

  NOTE: It will probably be necessary to open the test header control valve in the pump room.

- Location of spare or replacement sprinklers

- Name of building owner or official to contact in an emergency

- A list of sprinkler system deficiencies

- Location of all sprinkler system risers and the location of fire alarm control units needed to silence and reset the fire alarm system

- Design criteria as shown on the hydraulic nameplate and whether or not the system is a pipe schedule system

Fireground Operation Guidelines

Although individual fire department SOPs may vary, one procedure should be the same for all departments: Upon arrival at a sprinklered property, immediate preparations should be made to supply the fire department connection (FDC). The first-in engine company should locate the FDC and the nearest suitable fire hydrant. If there is any indication of an actual fire, such as smoke showing or the ringing of a sprinkler alarm, a minimum of two 2½-inch (65 mm) hoselines or one 3-inch (77 mm) hoseline should be connected to the FDC. Additionally, the engine company should lay a supply line(s) to the hydrant and make all the appropriate connections.

NOTE: Some fire departments use the second-in engine to connect to the FDC.

The interior attack crews should locate the fire and determine whether charging of the sprinkler system is necessary. It may be argued that it is prudent to charge the system immediately upon arrival of the fire companies. Obviously, this action is appropriate if a fire is evident. In some situations, however, the sprinklers may have extinguished the fire, or the system activation may be a malfunction. In most cases, it is desirable to confirm the presence of fire before pumping into the system (keeping in mind that water damage usually causes more damage than the fire itself).

Should the system need to be charged, the pump operator should slowly develop 150 psi (1 050 kPa) at the engine and maintain this pressure. In many situations, 150 psi (1 050 kPa) may be much more than is needed; however, it is a standard rule of thumb for supply at the FDC. From outside the building, it is usually difficult to determine how many sprinklers are operating in the fire area and the amount of friction loss in the system as a result. If it becomes obvious that the fire is spreading, additional lines should be connected to the fire department connection and charged. A rapid size-up may be hindered by low visibility because of the smoke being cooled by the sprinklers. This cooling causes the smoke to lose its normal buoyancy.

A firefighter familiar with the building should immediately check the control valves, if they are accessible, to ensure that they are open. Closed valves should be opened except when it is known that the building or area has been undergoing construction or renovation affecting the sprinklers. Opening valves under these circumstances would cause a severe loss of water to the system. The firefighter assigned to this task should carry a flashlight and a portable radio. If a pump supplies the system, the firefighter should also ensure that the pump is running. Control valves are frequently located in the pump room so that both functions can be performed by the same firefighter.

Firefighters should advance handlines as necessary. In an ideal situation, a properly designed sprinkler system will have controlled the fire, and only a small-diameter line should be necessary to complete extinguishment. However, there should be no hesitation in advancing larger or additional lines if the fire escalates. IFSTA recommends that handlines for interior structural fire fighting be not less than 1½-inch (38 mm) in diameter and preferably 1¾-inch (44.5 mm).

Through preincident planning, fire department personnel should have identified water supplies that will not lessen the amount of water available to the sprinkler or standpipe system. Firefighters should make sure that these water sources are the ones used when the actual incident is in progress. The design requirements for hydraulically calculated systems include an allowance for hose streams that are in addition to the sprinkler system demand. Flow rates for hose streams in occupancies are as follows:

- Light hazard = 100 gpm
- Ordinary hazard = 250 gpm
- Extra hazard = 500 gpm
- High-piled storage = 500 gpm

Therefore, these hose stream flow rates may be used without concern for depleting the water supply to the sprinkler system.

Truck companies should perform ventilation and salvage. The effectiveness of roof ventilation may be reduced by the cooling effect of the sprinkler discharge. In these cases, horizontal ventilation using smoke ejectors or blowers may be more effective. Performing horizontal ventilation often requires less work and personnel than would be needed to open a roof. This allows more firefighters to perform salvage operations.

There are situations where the combined efforts of sprinklers and fire companies are inadequate and a major fire develops. In these situations, partial or total structural collapse may occur. This collapse may result in broken sprinkler lines, resulting in a tremendous loss of water through broken lines. If the building is equipped with one or more outside post indicator valves, these valves can be closed to conserve water.

WARNING!
When fighting fires in a sprinklered building, DO NOT shut the sprinkler system off until the fire has been extinguished.

Many disastrous losses have occurred from shutting off sprinkler systems when the fire "appeared" to be under control. If the area involved in the fire is served by a sectional or floor control valve, that valve, rather than the main riser valve, should be closed to minimize damage to the area affected.

Occasionally during overhaul, hidden fire is uncovered and redevelops. For this reason, it is good practice for a firefighter with a portable radio to remain at the control valve until overhaul is completed.

The pump operator should also be ordered to shut down lines to the FDC. The piping from the fire department connection may bypass the main valve. In the absence of a sectional valve, water will continue to flow until the pumper is shut down.

Fire crews must give high priority to salvage operations during fire fighting operations and after the fire has been extinguished. In addition to the products of combustion, a large quantity of water is introduced to the area when sprinklers activate. All contents in the area of operating sprinklers should be covered with salvage covers. Particular attention should be given to the floor below the one on which the sprinklers are operating. Water may seep down several levels. All contents should be covered, and any objects on the floor that may be damaged by water should be raised.

The incident commander must be aware of the weight of the water discharged from the sprinkler system and hose streams, particularly where commodities have a high absorption capability such as in paper or clothing storage facilities. The added weight could result in structural collapse of the building. For more information on salvage operations, see IFSTA's ***Structural Fire Fighting: Truck Company Skills and Tactics*** manual.

Until the officer in charge has determined that a fire is sufficiently extinguished, the control valve must not be closed. In order to decrease water damage, specially designed devices can be used to shut off the flow of water at the operating sprinklers. Firefighters can carry several types of sprinkler tongs and wedges in the pockets of their turnout coats. Other types of tongs are attached to poles, which eliminates the need for a ladder. A wedge is made of soft wood and is designed to be driven into position with the heel of the hand until it is wedged between the sprinkler frame and the orifice. If the wedge is properly made, practically all of the water flow can be stopped. Sprinkler tongs are more effective in stopping the water flow because the rubber or neoprene seal of the tong eliminates dripping when applied properly.

Fireground Operations in One- and Two-Family Dwellings Equipped with Residential Fire Sprinklers

Residential sprinkler systems are designed to provide a higher level of life safety than is provided by smoke detectors and building materials in one- and two- family dwellings. Residential systems are less complex than commercial fire sprinkler systems, yet have unique features that must be taken into consideration when working in dwellings where they are installed.

Most fire departments have their own SOPs for residential structure fires. The same procedures that are normally followed by a department during suppression activities in nonsprinklered dwellings should be followed in a fire involving a residence with fire sprinklers. The main differences between a commercial occupancy having fire sprinklers and a dwelling with a residential fire sprinkler include the following:

- There is no fire department connection to boost the sprinkler system pressure and flow.

- Sprinkler piping is usually plastic and is not as strong as the piping installed in a commercial fire sprinkler system.

- Restoring the system is usually much easier following an activation than restoring a commercial fire sprinkler system.

- Residential fire sprinkler systems are most often a branch water line connected to the residential service pipe.

Residential sprinklers are designed for life safety and fire control, not property conservation. Because dwellings typically have a relatively low fire load, this type of sprinkler system uses much less water than a commercial system. A typical residential fire sprinkler system is designed to deliver from 13 to 23 gpm (52 L/min to 92 L/min) for a period of 10 minutes. There is no fire department connection (FDC) in the front of a private residence and no way to support the sprinkler system directly. Additional water needs are calculated in the original size-up and are provided as needed by the engine companies advancing suppression hoselines into the dwelling.

Most residential fire sprinklers, depending upon local water authority requirements, are connected directly to the residential water service line and not directly to a water main. NFPA® 13D allows several different configurations for water supply design to be employed with these systems. Local water authority requirements and regulations should be determined and then compared with those configurations outlined in NFPA® 13D prior to installation.

NFPA® 13D allows a considerable amount of leeway in the integrity of sprinkler piping in residential systems. These systems are designed to be affordable for the average homeowner. Therefore, industrial-strength piping is not necessary. This is important to remember during ventilation and overhaul operations. Residential sprinkler piping is often designed for CPVC plastic piping, copper piping, or thin-wall steel piping installation. Each is susceptible to damage during ordinary fire fighting operations. While the sprinkler system can be shut down, this option may not be advisable, especially if the fire is not yet fully extinguished. A preincident familiarization is recommended for all sprinklered buildings, which includes residential buildings.

System restoration is simply a matter of replacing the activated sprinkler(s) and turning the sprinkler valve back on. Although an alarm device may be installed, it will not be the type that has to be reset in order to place the system back in service.